Food Insecurity

This book explores the experiences, causes, and consequences of food insecurity in different geographical regions and historical eras. It highlights collective and political actions aimed at food sovereignty as solutions to mitigate suffering.

Despite global efforts to end hunger, it persists and has even increased in some regions. This book provides interdisciplinary and historical perspectives on the manifestations of food insecurity, with case studies illustrating how people coped with violations of their rights during war-time deprivation in France; the neoliberal incursions on food supply in Turkey, Greece, and Nicaragua; as well as the consequences of radioactive contamination of farmland in Japan. This edited collection adopts an analytical approach to understanding food insecurity by examining how the historical and political situations in different countries have resulted in an unfolding dialectic of food insecurity and resistance, with the most marginalized people—immigrants, those in refugee camps, poor peasants, and so forth—consistently suffering the worst effects, yet still maintaining agency to fight back.

The book tackles food insecurity on a local as well as a global scale and will thus be useful for a broad range of audiences, including students, scholars, and the general public interested in studying food crises, globalization, and current global issues.

Tamar Mayer is the Robert R. Churchill Professor of Geosciences at Middlebury College, Vermont, where she is the director of the Rohatyn Center for Global Affairs. She is the editor or co-editor of five books that focus on various dimensions of international and global crises.

Molly D. Anderson is the William R. Kenan Jr. Professor of Food Studies and Academic Director of Food Studies at Middlebury College, Vermont. She works on food system transformations toward greater resilience and sustainability, the right to food, and the intersections of civil society and academic perspectives.

Critical Food Studies

Series editors: Michael K. Goodman, University of Reading, UK, and Colin Sage, Independent Scholar

The study of food has seldom been more pressing or prescient. From the intensifying globalization of food, a world-wide food crisis, and the continuing inequalities of its production and consumption, to food's exploding media presence, and its growing re-connections to places and people through 'alternative food movements', this series promotes critical explorations of contemporary food cultures and politics. Building on previous but disparate scholarship, its overall aims are to develop innovative and theoretical lenses and empirical material in order to contribute to – but also begin to more fully delineate – the confines and confluences of an agenda of critical food research and writing.

Of particular concern are original theoretical and empirical treatments of the materializations of food politics, meanings and representations, the shifting political economies and ecologies of food production and consumption, and the growing transgressions between alternative and corporatist food networks.

Alternative Food Politics
From the Margins to the Mainstream
Edited by Michelle Phillipov and Katherine Kirkwood

Hunger and Postcolonial Writing
Muzna Rahman

Digital Food Cultures
Edited by Deborah Lupton and Zeena Feldman

Food Insecurity
A Matter of Justice, Sovereignty, and Survival
Edited by Tamar Mayer and Molly D. Anderson

For more information about this series, please visit: www.routledge.com/Critical-Food-Studies/book-series/CFS

Food Insecurity

A Matter of Justice, Sovereignty, and Survival

Edited by
Tamar Mayer and Molly D. Anderson

Routledge
Taylor & Francis Group
LONDON AND NEW YORK

First published 2021
by Routledge
2 Park Square, Milton Park, Abingdon, Oxon OX14 4RN

and by Routledge
605 Third Avenue, New York, NY 10017

First issued in paperback 2022

Routledge is an imprint of the Taylor & Francis Group, an informa business

Publisher's Note
The publisher has gone to great lengths to ensure the quality of this reprint but points out that some imperfections in the original copies may be apparent.

British Library Cataloguing-in-Publication Data
A catalogue record for this book is available from the British Library

Library of Congress Cataloging-in-Publication Data
A catalog record has been requested for this book

ISBN: 978-1-138-35885-0 (hbk)
ISBN: 978-0-367-50615-5 (pbk)
ISBN: 978-0-429-43408-2 (ebk)

DOI: 10.4324/9780429434082

Typeset in Times New Roman
by Newgen Publishing UK

Contents

Figures

Tables

Contributors

Molly D. Anderson teaches about food security, food justice, and food system transitions at Middlebury College in Vermont and is developing a new food studies program. She is involved in food system planning at the state and national scales, participating in the Vermont Farm to Plate Network and the Inter-Institutional Network for Food, Agriculture & Sustainability; and she is a member of the International Panel of Experts on Sustainable Food Systems (IPES-Food). She has worked for Oxfam America and Tufts University, where she was the founding director of the Agriculture, Food and Environment Graduate Program and directed Tufts Institute of the Environment.

Lindsey Carte is an assistant professor in the Núcleo de Ciencias Sociales y Humanidades at the Universidad de la Frontera, Temuco, Chile. Her research interests include human migration, rural change and gender in Latin America. Lindsey's recent projects examine the experiences of Central American immigrants to Mexico and linkages between rural and environmental change and migration in Guatemala and Nicaragua. She currently engages in research on the expansion of forest plantations and migration in Chile.

Jampel Dell'Angelo is an assistant professor of water governance at the Vrije Universiteit of Amsterdam. He conducts research at the Institute of Environmental Studies (IVM) and is a visiting fellow at the Department of Environmental Science, Policy, and Management (ESPM) of the University of California, Berkeley, and an affiliate researcher at the National Socio-Environmental Synthesis Center (SESYNC), Annapolis, Maryland. Dell'Angelo is an environmental social scientist with a background in political economy and political ecology of natural resources. His research, mainly focused on competition and cooperation over water resources, spans different levels, from river basin to global dynamics, and several countries.

Paolo D'Odorico is a professor of hydrology at the Department of Environmental Science, Policy, Management (ESPM) at the University of California, Berkeley. He has research expertise in Ecohydrology, Surface Hydrology, Ecosystem Ecology, Aeolian Processes, Desertification, and

Water and Food Security, with over two hundred publications in these topics. Professor D'Odorico is currently investigating the globalization of water through virtual water trade and international land investments, and their impact on water equity, societal resilience, environmental stewardship, and food security.

Pierre Fuller is the author of *Famine Relief in Warlord China* (Harvard University Asia Center, 2019). In 2015–16 he was a research fellow at the Shelby Cullom Davis Center for Historical Studies at Princeton University, in New Jersey, where he began his current book exploring the role that cultural memory of natural disasters has played in revolutionary politics in modern China.

Adam Gilbertson is an associate research scientist at the Pacific Institute for Research and Evaluation, Chapel Hill Center, Chapel Hill, North Carolina. His doctoral research considered the relationship between food insecurity, everyday food practices, and experiences of intimate partner violence within Kenyan informal settlements. Currently, his work focuses on the social and ethical implications of voluntary medical male circumcision for HIV prevention in eastern and southern Africa and food security among older people living with HIV in the United States.

Faik Gür is an associate professor of development studies at the Department of International Relations at Ozyegin University, Istanbul. He regularly teaches courses and conducts research on the sociology of modernization, nations and nationalism, contemporary political geography, agrarian transformation, and critical food studies. Faik Gür has published several articles in international and national journals. Recently he co-edited a book titled *From the Perspective of Sustainable Life: Local and Rural Development,* published by Ozyegin University Press (2018).

Kyle Harper is a historian of the classical world and Senior Vice President and Provost at the University of Oklahoma. His first book was *Slavery in the Late Roman World, AD 275–425* (2011). The book was awarded the James Henry Breasted Prize by the American Historical Association. His second book, *From Shame to Sin: The Christian Transformation of Sexual Morality*, published in 2013, received the Award for Excellence in Historical Studies from the American Academy of Religion. His third book, *The Fate of Rome: Climate, Disease, and the End of an Empire*, appeared in 2017.

Joost Jongerden is an associate professor at Rural Sociology, Wageningen University, the Netherlands, and project professor at the Asian Platform for Global Sustainability & Transcultural Studies at Kyoto University, Japan. A common denominator of his research has been the question of how people create and maintain a livable life under conditions of precarity. This interest has expressed itself in the ways in which people develop alternatives for market- and state-induced insecurities. This he refers to as 'Do-It-Yourself-Development'. A list of his publications is available at https://joostjongerden.academia.edu/.

Joo-Young Jung is a professor of media and communication at International Christian University in Tokyo. Her research interests include dynamic relationships of new and old media in the context of particular spatial, social, and communication environments. She has examined communication processes in different spatial contexts, such as urban communities in different cities; in particular social contexts, such as disaster and post-disaster situations; and in changing communication environments. Her research has been published in major communication journals and books, including her recent co-edited book titled, *The Communication Ecology of 21st Century Urban Communities* (2018).

Charalampos (Harry) Konstantinidis is an associate professor of economics at the University of Massachusetts, Boston. His main research interests are political economy, ecological economics, and the political economy of food and agriculture, with a particular emphasis on Europe. His recent work examines the transformations of the agri-food systems and the undermining of the right to food during the economic adjustment programs in Greece. His work has appeared, among others, in the *Review of Social Economy*, *Feminist Economics*, and the *Review of Radical Political Economics.*

Teresa Mares is an associate professor of anthropology at the University of Vermont. She is the author of *Life on the Other Border: Farmworkers and Food Justice in Vermont* (2019). Her work has appeared in several journals, including *Food and Foodways*, *Latino Studies*, *Agriculture and Human Values*, *Working USA*, *Journal of Pedagogic Development*, and the *Journal of Agricultural, Food Systems and Community Development*, in addition to numerous edited volumes. She is the co-director of Huertas, a food security project for Latino/a dairy farmworkers connected to UVM Extension's Bridges to Health Program.

Tamar Mayer is the Robert R. Churchill Professor of Geosciences at Middlebury College, Vermont, where she is director of both the Rohatyn Center for Global Affairs and the Program in International and Global Studies. She is a political geographer whose research interests lie in the interplay among nationalism, landscape and memory as it pertains to stateless ethnic nations, especially in the Middle East and Western China. She has edited or co-edited five books that focus on different dimensions of international and global issues, the most recent of which is *The Crisis in Global Youth Unemployment* (2019).

Jessie Mazar received her MS in Food Systems in 2016 from the University of Vermont. She is an associate faculty at Feather River College, where she teaches for the Ecological Farming Certificate program and maintains a small farm. Jessie worked on an applied research project in Vermont from 2011 to 2017 that investigated food-access strategies among Latino/a migrant farm workers. She co-facilitated a community-based food-access project, called Huertas, that helps to plant kitchen gardens with fresh, culturally familiar vegetables and herbs on farms where farm workers are living and working. She received support through a Northeast Sustainable Agriculture Research and Education (NE-SARE) grant and was awarded the Thomas Marchione Food as a Human Right Prize from the Society of Anthropology of Food and Nutrition.

Santana Navarro Olmedo is a researcher at Ecosur (El Colegio de la Frontera Sur) in Mexico and holds a PhD in Ecology and Sustainable Development and a master's degree in Social Anthropology. His research bridges these disciplines through its focus on three major thematic axes: rural development, socio-environmental conflicts, and public policy and intercultural education. Specifically, his publications have focused on the analysis of conditional transfer programs and on socio-environmental conflicts in communities of the southern Yucatan Peninsula, as well as community development and intercultural relations in the state of Oaxaca. He has worked closely with local communities in Mexico's Calakmul region since 2011.

Ellen Oxfeld is a professor of anthropology at Middlebury College, Vermont. She is the author of a recent ethnography of food culture in rural southeastern China entitled *Bitter and Sweet: Food, Meaning and Modernity in Rural China* (2017). She is also the author of *Blood, Sweat and Mahjong: Family and Enterprise in an Overseas Chinese Community* (1993), and *"Drink Water, but Remember the Source": Moral Discourse in a Chinese Village* (2010). She is co-editor, along with Lynellyn Long, of *Coming Home? Refugees, Immigrants and Those Who Stayed Behind* (2004). She is currently working on a new project, exploring meanings of commensality in contemporary China.

Murat Öztürk is an associate professor of development economics at the Department of Economics at Kırklareli University in Turkey. He regularly teaches and conducts research on agricultural economics, political economy, and international political economy. He is author of the book, *Agriculture, Peasantry and Poverty in Turkey in the Neo-liberal Age* (2012). He is also co-author of the chapter, "Commodification and Social Commons: Smallholder Autonomy and 'Rurban' Relations in Turkey," in *Perspectives on Commoning* (2017).

Claudia Radel is a professor of human geography in the department of Environment and Society, Utah State University, and associate dean of the College of Natural Resources. Her research focuses at the intersection of natural resources and social inequalities, with an emphasis on understanding processes impacting the wellbeing and social relations of smallholder farming families in Latin America. Her research is funded by the U.S. National Science Foundation and has appeared in a wide range of journals, such as *World Development*, *Human Ecology*, and *Journal of Agrarian Change*. She has a PhD in Geography and an MPA in International Development.

Maria Cristina Rulli is an associate professor in hydrology and water engineering at the Politecnico di Milano, Italy. She is an expert in hydrological modeling and water and food security. Rulli applies quantitative methods to investigate local and global complex sustainability problems that range from land and water grabbing, the food–energy–water nexus, to deforestation and the spread of global pandemics. She is currently investigating the direct effects of hydrologic extremes (e.g., floods, droughts, landslides, hurricanes, typhoons) on food production and availability and the cascades effects on conflicts and human migrations occurrence. Rulli has won several international awards. and her work has been published in leading journals.

Birgit Schmook holds a master's degree in agriculture and a doctorate in geography. Her overall research interests are in the field of human-environment relationship. Throughout her 24-year tenure at Ecosur (El Colegio de la Frontera Sur), Mexico, she has dedicated substantial time and effort to participatory research. She is a full professor and has directed and participated in various research projects on topics such as the impact of migration on local farming systems, land-use changes, and more recently food security and land-tenure regimes in Mexico and Nicaragua. She teaches courses on poverty and sustainable development, as well as a field course on tropical ecosystems. She has also been an adjunct professor at Utah State University since 2014.

Paula Schwartz is the Lois B. Watson Professor of French Studies at Middlebury College, Vermont, where she teaches courses on twentieth-century French history and culture. Her scholarship focuses on the French Resistance movement of the Second World War, food studies and food protests, gender studies, and French communism. Her book, *Today Sardines Are Not for Sale: A Street Protest in Occupied Paris* (2020), is a microhistory of a demonstration for food amidst wartime shortages.

Patricia A. Stapleton is a comparative political science and public policy scholar. She holds an assistant professor position in the Society, Technology, and Policy Program at Worcester Polytechnic Institute in Massachusetts. Her research interests cover the regulation of biotechnology, both in food and reproductive medicine. In addition to a PhD in political science, she has an MA in French literature. Occasionally, she also writes on the intersection of literature and politics, especially in the context of utopian/dystopian fiction.

Tomiko Yamaguchi is a professor of sociology at International Christian University in Tokyo. Her study focus lies at the intersection of the sociology of agriculture and food and science and technology studies. She is particularly interested in the value conflicts surrounding the use of novel foods and discursive battles about the meaning of safety, examined cases—including GMOs, radioactive nuclides, and gene editing technologies. She has published in numerous journals and books, including *Science, Technology and Human Values* and *Food Policy*. Her most recent publication is *Simulation, Prediction and Society: The Politics of Forecasting* (published in Japanese, 2019).

Acknowledgments

The work on this book began with a conference titled, *Food Insecurity in a Globalized World: The Politics and Culture of Food Systems*, which we organized with our colleague, Professor Paula Schwarz, at Middlebury College, in 2016. This conference was the fourth international and interdisciplinary conference on a global theme, hosted annually by Middlebury's Rohatyn Center for Global Affairs. We want to thank the many departments at the College and the Rohatyn Center for Global Affairs for their generous support, without which we would not have been able to assemble an accomplished group of scholars from different part of the world. We owe our gratitude to the staff of the Center, who helped make the conference a success. Most of the chapters in this volume were presented at the conference, but we solicited additional essays on topics and geographical areas that were not addressed at the 2016 conference. We thank our student research assistants Laura Dillon, Kaleb Patterson, Elizabeth Sawyer, Greg Treiman, and Suria Vanrajah. We are also thankful to Mary Bagg, our copyeditor. Lastly, we want to thank our editors at Routledge: Mike Goodman, Colin Sage, Ruth Anderson, Faye Leerink and Nonita Saha for their support and patience.

Tamar Mayer and Molly D. Anderson

1 Food insecurity in context

Molly D. Anderson and Tamar Mayer

> After decades of steady decline, the trend in world hunger … reverted in 2015, remaining virtually unchanged in the past three years at a level slightly below 11 percent. Meanwhile, the number of people who suffer from hunger has slowly increased. As a result, more than 820 million people in the world were still hungry in 2018, underscoring the immense challenge of achieving the Zero Hunger target by 2030.
>
> (FAO, IFAD, UNICEF, WFP, and WHO 2019: xiv)

Despite the goal to end hunger in the world, as set forth in 2015 in the UN Sustainable Development Goals, hunger remains a significant problem worldwide. Severe food insecurity was worse in 2018 than it was in 2014 in every region of the world except North America and Europe. While Middle and Eastern Africa have the highest prevalence of undernourishment (26.5% and 30.8% respectively), Asia has the largest number of chronically malnourished people; and the situation is worsening in Latin America, the Caribbean, other regions of Africa, and Western Asia. The number of undernourished people in sub-Saharan Africa rose from 181 million in 2010 to 239 million in 2018, and in Africa as a continent, almost 20 percent of the population (more than 256 million) is chronically undernourished. The decreasing trend in undernourishment that characterized Asia until recently seems to be slowing down significantly; 514 million people there were chronically undernourished in 2018. Several countries in Southeast Asia have been affected by climate conditions that have worsened food availability and prices, while countries in Western Asia have been affected by prolonged armed conflicts. Countries encountering economic downturns and slowdowns have also shown worsening hunger and food insecurity (Food and Agriculture Organization (FAO), International Fund for Agricultural Development (IFAD), United Nations Children's Fund (UNICEF), World Food Program (WFP), and World Health Organization (WHO) 2019).

The primary metric used to estimate severe food insecurity in the annual State of Food Insecurity and Nutrition in the World (SOFI; previously the State of Food Insecurity in the World) is Prevalence of Undernourishment

(PoU), although serious issues with its accuracy have been identified by several scholars (Lappé et al. 2013). PoU has been supplemented in the more recent SOFIs with five other metrics, including the Food Insecurity Experience Scale (FIES), the first globally applied self-reported metric of food insecurity. According to the FIES, moderate or severe food insecurity was reported by 57.7 percent of surveyed respondents in sub-Saharan Africa in 2018, contrasting with a PoU of 22.8 percent. If health problems associated with micronutrient deficiencies, obesity, and overweight are included in the tally, over one-third of the world's population is inadequately nourished. The global number of obese people surpassed the global number of undernourished people in 2016. In 2018, childhood overweight affected 40.1 million children under five worldwide; and in 2016, nearly two in five adults (38.9 percent) were overweight, representing 2 billion adults worldwide. The prevalence of overweight is increasing in all age groups and in all regions (FAO, IFAD, UNICEF, WFP, and WHO 2019).

Poor access to healthy food affects all age groups, but particularly women and children. They cannot develop to their full mental and physical potential, and they face increased threats from disease, a decreased quality of life, and shortened lifespans. Mothers who are food insecure during pregnancy can pass along epigenetic metabolic markers that result in obesity and poor health for their children; early malnutrition in children has similar results (Landeker 2011). Compromised access to food often results in tensions within families; and as some adults migrate in search of better livelihoods, communities further erode, and families are negatively affected. As global regimes determine the agricultural landscape and new geopolitics of food are established, old food cultures disappear, and communities plant fewer traditional crops, with negative consequences for their nutritional status.

We need to ask why it is that, despite all the efforts made to date, there has been so little progress, and why at least one out of every nine people in the world is still undernourished. Questions of *who* eats *what, when, where*, and *why* are at the center of this volume. The chapters in this volume establish that food insecurity is a matter of justice and survival, regardless of geography and history. They analyze from multiple disciplinary perspectives the experiences of, and responses to, food insecurity at different geographical scales, from the local to the global. Several chapters demonstrate how *food sovereignty* is emerging in different places as a form of resistance to the persistent injustice of food insecurity.

Justice and survival

Deep-rooted political and economic theories underpin concepts of *food justice*, which in turn underpin countries' decisions about how to address food insecurity. The lack of food justice is most apparent in the stark discrepancy in food security and very low food security between countries and regions. Yet every country, even the wealthiest, has populations that suffer from food insecurity. The questions about food justice and its troubling persistence are related to questions about social good that go back to Plato's *Republic*, in which he tried to identify the ideals that would lead to a just society.

One of the long-standing debates about justice, with implications for food justice, is whether some people have a right to set the standards of social justice for all, or whether each individual has the freedom to hold alternative views. These debates played out in the writings of Jeremy Bentham and John Stuart Mill, who laid the groundwork for utilitarian political philosophy, and were extended by Adam Smith to the view that the market could best determine the greatest good for all (DuPuis et al. 2013). For others, a market in which each person was driven by individual interest seemed to be the making of a *Lord of the Flies* society, anticipated early on by Thomas Hobbes in his arguments for the necessity of hierarchies and a social contract. Other philosophers and diplomats developed universal standards, such as the US Declaration of Independence and the UN Universal Declaration of Human Rights. Of course, the "universal" standards that were expressed in high-sounding prose in fact applied, in their enactment, only to a subset of humanity. For example, the Declaration of Independence left out women, slaves, and indigenous peoples.

In *A Theory of Justice*, John Rawls (1971) advanced the attempt to reconcile competing views of individual liberty and social order created through a social contract. He proposed that justice could be determined by a "veil of ignorance," such that each individual would choose a society in which the most marginalized people would be served with equity, were they ignorant of the consequences of their choices on themselves. Thus, people would accept inequitable distribution of wealth if they knew that it would benefit people who had the least, given that they might be among "the least" once the veil was lifted.

Melanie DuPuis and colleagues (2013) have suggested that three main perspectives critique the Rawlsian ideal—political economy, communitarian, and cultural—and draw the lines to food justice. They trace the political economy critique to Karl Marx, who saw economic goods as a "fetish" or illusory form of freedom. For Marx, the exploitative relationships of labor and capital are the base of social injustice, and social activists must fight for a radical restructuring of society that allows labor to control the means of production. Communitarians believe in a set of social and cultural values that are widely shared and form the basis of a moral economy. Michael Walzer (1990) and Michael Sandel (1984) are two contemporary communitarians, arguing that communities should have the freedom to establish their own values and rules. These perspectives are shared by indigenous communities and other food sovereignty advocates, who believe that each community (however defined) has a sovereignty that must be respected.

DuPuis and colleagues (2013) go on to describe the cultural perspective as held by feminist, postcolonial, and critical race theorists. Their arguments stand against the political economy theories and communitarian theories in claiming that both universalist and particularist perspectives on justice place the Western white male citizen as the "universal," and that different groups with different histories of oppression must be able to form their own ideals of the good life and social justice. While DuPuis and colleagues explore how these different perspectives on social justice are manifest in arguments about localism and local food, they have wider connections with food justice.

Views on the distribution of food, one of the key actions associated with food justice, often depend on which theory of food justice one holds. Distributive-justice theories have tended to regard food as just another of the goods to be distributed (Scoville 2015). But food is quite different from cell phones or automobiles; its absence is a literal matter of survival, and its presence is a matter of identity as well as nutrition and health. This can justifiably lead to very different treatment in every agreement from international trade (hence La Via Campesina's cry to "get agriculture out of the WTO") to household standards on who eats first when food is limited (Rosset 2006).

While distributive justice has been a primary focus of food justice theory, participatory justice is also extremely important (Dieterle 2015; Moragues-Faus 2017). That is, it matters not only whether people get food but also whether they have agency to advocate for themselves (with positive effects). Ana Moragues-Faus demonstrates, in analyzing narratives of food security, that political and global elements of who is a subject of food injustice (and why) have been largely bypassed in debates about food security. How food security is framed in governmental and intergovernmental forums has significant impacts on the solutions that are adopted; when systemic inequities in political, cultural, and socioeconomic domains are understood as the main cause of food insecurity, the rights of poor and marginalized people to participate in the determination of solutions can be recognized.

All the chapters in this volume make the underlying assumption that food is a right, and that lack of access to safe and healthy food, regardless of reason, is directly related to lack of justice. The right to food has expanded to address healthy food, not just calories. What this right means in practice has been illuminated by several decades of scholarship and publications of the Committee on Economic, Social and Cultural Rights, FAO's Right to Food Unit, and the UN Special Rapporteur on the Right to Food. Olivier de Schutter, UN Special Rapporteur from 2008 to 2014, was especially forceful and influential in articulating a clear critique of industrial agriculture and expanding the right to food to encompass nutritional security (Sage 2014). The right to food entails procedural approaches, such as full participation by those whose rights have been violated and regular monitoring of progress toward respecting, protecting, and fulfilling the right. (These approaches are detailed in the Voluntary Guidelines on the Progressive Realization of the Right to Adequate Food in the Context of National Food Security, developed by the Committee on World Food Security and adopted by the FAO Council in 2004.) While the right to food and nutrition is a comprehensive scaffolding for achieving food security, scholars turned to the related concepts of food justice and food sovereignty as the failures of governmental and philanthropic attempts to increase food security became more apparent.

Themes that have emerged, particularly since the 1990s, have included a shift from focusing on consumers to focusing on workers, the class- and race-based causes of food injustice, and global manifestations (Herman and Goodman 2018). The specific ways in which populations working within certain sectors of the economy experience food injustice have entered public awareness. The plight of farmworkers, whose exploited labor props up the industrialized food system, was exposed to the US public by authors such as Eric Schlosser (2001) and organizations such as the Coalition of Immokalee Workers, which won seven cases demonstrating enslavement of farmworkers in Florida and launched a successful campaign for "fair food." Similar abuses were revealed against people catching and processing fish, processing livestock, and working in retail and restaurant industries (McMillan 2012; Genoways 2014; Anderson and Athreya 2015; Food Chain Workers Alliance and Solidarity Cooperative 2016; Environmental Justice Foundation 2019). In the United States and Canada, attention turned to "food apartheid" or the exclusion of people of color from access to healthy food, in addition to a renewed focus on the theft of farmland from black farmers (Gilbert et al. 2002; Alkon and Agyeman 2011; Sbicca 2012; Newkirk 2019). While injustices suffered by African Americans as both producers and consumers of food have received overdue attention (e.g., a cover story in the September 2019 issue of the magazine, *The Atlantic*), the suffering of indigenous people from land theft, forced removal to inhospitable reservations, lack of access to healthy food, and astronomical rates of diet-related diseases such as diabetes has not yet gotten the public awareness and attention to possible reparations that it deserves (Wiedman 2012).

At a more theoretical level, political economy frameworks and the study of social movements allowed analysts to better understand the role of power and inequality in the food system (Bonanno and Wolf 2018; Sbicca 2018; Anderson et al. 2019). Similarly, focusing on social determinants of health allowed nutritionists and public health experts to comprehend how inequality, poverty, and intersectionality of different forms of discrimination interact in the path from diet to poor health. Inequality between classes and nations largely explains why some people have more than enough to eat while others starve (Patel 2008), although obesity and overweight often coexist with hunger and other forms of undernourishment. The 2019 SOFI examines this paradox, first highlighted in recognition of the "nutrition transition" (Popkin 1993) whereby economic development is accompanied by rising access to cheap foods high in sugar, fat, and salt. This is not a necessary conjunction: corporate food manufacturers, agribusiness, and drug manufacturers have encouraged the overuse of products resulting in malnutrition and food system related health problems, ranging from rising antibiotic resistance due to excessive application in livestock operations to pesticide exposure to obesity and its correlates (Nestle 2002; Patel 2008; IPES-Food 2017).

Food injustice can be a lens for the critique of capitalism, in particular the neoliberal form of capitalism that has spread since the 1980s. Understanding food injustice requires understanding how capitalism works (Holt-Giménez 2017; Clapp and Isakson 2018). This critique leads to imagining post-capitalist food systems that might be more equitable and just. "Commoning" is one such system, in which food and the risks of food production are distributed more equitably through social relations that recognize values other than money (Vivero-Pol 2019). Food sharing is another example of emerging post-capitalist food distribution systems (Carolan 2017), although of course its roots are deep in every culture.

As long as people who hold political and economic power have access to safe food while the majority of their fellow citizens do not, and as long as powerful states have better access to food than others, justice cannot be served. Further, despite signing agreements to uphold the right to food and nutrition, in reality the practice is not universal. In recognition of violations of rights and structural problems that prevent access to healthy food, there has been a demand for greater localized control over one's food system—in other words, food sovereignty.

Food sovereignty and survival

Food sovereignty is a form of resistance against the multiple inequities in access to resources and power that lead to food insecurity. It is a model that seeks to democratize the food system to benefit small-scale farmers and promote sustainable production (Dekeyser et al. 2018). The concept originated in Mexico in the 1980s (Edelman 2014a), but it has also been attributed to the peasant organization La Via Campesina (Desmarais 2007). In 2007, more than five hundred representatives from eighty countries came together in the village of Nyéléni in Mali as a global movement for food sovereignty. The movement includes peasants and family farmers; artisanal fisherfolk; indigenous peoples; landless people; rural workers; migrants; pastoralists; forest communities; women; youth; and consumer, environmental, and urban movements. The Declaration of Nyéléni, written at this gathering, described the ability of these food producers to provide "enough healthy, good and abundant food" for the hungry world; it also articulated the threat to that capacity by neoliberalism and global capitalism (Nyéléni Forum 2007).

An important outcome of the Nyéléni gathering was the articulation of six pillars of food sovereignty: (1) a focus on providing sufficient, healthy, and culturally appropriate food for all individuals, peoples, and communities; (2) valuing food providers and respecting their rights; (3) localizing food systems and bringing providers and consumers to the center of decision-making about food issues; (4) putting control over territory, land, grazing, water, seeds, livestock, and fish locally in the hands of food providers; (5) building knowledge and skills of food providers; and (6) working with nature in diverse, low-external-input agroecological production systems.

Food sovereignty, then, seeks to heal the economic, environmental, social, and cultural damage of industrialized food systems and opposes privatization of food systems and commodification of food. Thus, food sovereignty has become the dominant framework of resisting the international market-based food system that manages global food distribution from places where it can be produced cheaply (and often with lax labor and environmental regulations) to places where it can be sold at the highest price. Although "cheap food" is one of the goals of the US food system, it is only cheap when its environmental and social costs are externalized onto future generations, poor people within the country, or poor people in other countries.

While the vision for food sovereignty as a way to provide access to sufficient healthy and culturally appropriate food has been important as a framework of goals, its greater value has been the mobilization of different broad-based movements into a larger grassroots movement that strives to achieve the ideas developed in Nyéléni and in subsequent efforts. Food sovereignty has become the clarion call of social movements around the world, bringing together movements fighting for women's rights, peasants' rights, land and secure tenure, seeds, and better access to healthy food. Diverse movements joined in fighting for the UN Declaration on the Rights of Peasants and Others Working in Rural Areas, adopted in 2018.

Food sovereignty, while ostensibly a claim of strength and growing in support, might be seen as an admission of vulnerability. It is never used as the justification for taking away land, water, or other resources needed for food production from a weaker nation, tribe, or community. Rather, it is the accusation by a more vulnerable entity, and by interests that stand in solidarity with them, that goods and rights have been stolen. Despite the increased agency that some farmers and indigenous people around the world now have, food sovereignty has not been fully achieved anywhere yet. Beyond being an anti-colonial and anti-neoliberal critique, food sovereignty is not a blueprint of how to achieve the goals; and numerous questions remain about its meaning and implementation (Edelman et al. 2014).

Much of the recent scholarship on food sovereignty has attempted to explain how it differs from, or is complementary to, the concept of food security (e.g., Clapp 2014; Edelman 2014b; Jarocz 2014). Others have used participatory research methods to analyze the ways that communities have implemented food sovereignty and what it means to them. A strong emphasis in this scholarship, especially the work from the Center on Agroecology, Water and Resilience at Coventry University, is respect for different ways of knowing and what this means for conducting research (Pimbert 2006; Pimbert et al. 2010; People's Knowledge Editorial Collective 2016; Levkoe et al. 2018). Further recent work has analyzed how different communities have implemented food sovereignty and greater self-determination in their quest for sustainable food systems (e.g., Baker et al. 2019). The critique of neoliberal capitalism, combined with predictions of future catastrophe in the world food system due to climate change, have led to increasing calls for deep transformation and not just reform of the global food system (e.g., Holt-Giménez and Li 2011).

Although, to some extent, the causes of hunger and food insecurity are all a failure of food sovereignty and the ability of people to determine their own future, there are many contributing structural problems. The chapters in this volume address such causes in different geographical settings, historical periods, and scales (from the local to the global). But it is important to remember that these scales are neither discrete nor independent of each other and are best described as nested. Therefore, in order to understand food insecurity at the most local scale, as analyzed in Chapters 2, 3, 6, 8, 10, 11, and 12 (by Gilbertson, Mazar and Mares, Schmook et al., Harper, Oxfeld, Schwartz, and Yamaguchi and Jung, respectively), we need to include a discussion of the causes of hunger at both the national and global levels; we need to look at the impact of global processes—such as colonialism, neoliberal policies, and global warming—on those who inhabit the local scale and suffer from food insecurity the most. Chapters 4, 5, 7, 9, and 13 (by Konstantinidis, Öztürk et al., Dell'Angelo et al., Fuller, and Stapleton, respectively) address food insecurity at scales greater than the local. It is also important to learn from the past and understand how national responses to hunger have changed over time. Pierre Fuller does this in Chapter 9, using China's responses to famines in the nineteenth and twentieth centuries to show how they depend on ideology and the availability of food at the national and local levels.

Causes and consequences of food insecurity

At heart, every instance of food insecurity stems from inequitable power relations. Amartya Sen (1999:16) famously declared that famine has never occurred in a functioning democracy. We could say with equal surety that food insecurity does not occur in societies in which power is shared equitably; democracy is a manifestation of equity, where the government is responsive to its people. Every case of food insecurity in this volume can be interpreted as resulting from unequal access to resources needed for food production, opportunity, or political voice (and often, from all of these together). Sufficient food is produced for everyone in the world to have adequate calories (Lappé and Collins 2015), at least at the current point in global warming. But how should this food be allocated? Is "the market" adequate to distribute food to places of scarcity, or does it need to be mediated by public policy, perhaps based on recognition of the human right to food and nutrition?

UN agencies primarily attributed the decline in reliable access to food from 2014 to 2017 to climate change and political conflicts resulting in warfare (FAO, IFAD, UNICEF, WFP, and WHO 2019). These issues have been compounded by the failure to fully recover from the 2007–08 economic crisis, neoliberal austerity policies, scarcity of essential natural resources and environmental degradation, speculation in agricultural markets, and increasing demand for biofuels by wealthy countries. These contributed to the "food price crisis" of 2008 (McMichael 2009; Wise and Murphy 2013), which was followed by another sharp rise in 2010. Each of these factors alone has had a severe impact on food prices and food security and, because they are intertwined, their impact has been devastating, particularly in countries where governments have not been responsive to immediate needs of their population.

Climate change

The effects of climate change on food insecurity should be called out explicitly because this is the most dangerous harbinger, largely caused by greenhouse-gas emissions from relatively wealthy countries yet borne mostly by populations in poor countries who have benefited little from burning fossil fuels. Global warming is already 1°C above the postindustrial average, and, if current emission rates continue, the world is on track to hit 1.5°C above that mark as soon as 2030. The impact of global warming on human and natural systems is already apparent, and some of the services that ocean and land ecosystems provide have already changed for the worse. Even if emissions drop drastically in the near future, continued effects are locked in by the amounts of greenhouse gases that are already in the atmosphere. The effects include severe droughts, floods, tropical cyclones and hurricanes, heat waves sufficient to kill livestock and humans, sea-level rise and saltwater intrusion in low-lying coastal areas, and loss of biodiversity, including fish species on which human populations depend.

Those likely to be affected most are disadvantaged and vulnerable populations, some indigenous peoples, and local communities dependent on agricultural or coastal livelihoods. Regions at disproportionately higher risk include Arctic ecosystems, dryland regions, small island developing states, and least-developed countries (Masson-Delmotte et al. 2018). Higher temperatures will lead to yield reductions in maize, rice, wheat, and soybeans, with each degree Celsius increase in global mean temperature reducing global average yields of wheat by 6 percent, rice by 3.2 percent, maize by 7.4 percent, and soybeans by 3.1 percent (Zhao et al. 2017). In addition, with higher CO_2 concentrations the quantities of available protein, iron, and zinc in wheat, rice, maize, and soybeans will drop; at least 2 billion people already suffer from iron and zinc deficiencies, and global warming will increase these numbers (Myers et al. 2014). The IPCC report on climate change and land highlights the urgency of food system transformation, changing practices at every stage from production through consumption and waste disposal (IPCC 2019a). It predicts sharp increases in global food insecurity if actions are not taken rapidly to mitigate climate change, as droughts, floods, and catastrophic storms take their toll. The IPCC report on oceans and the cryosphere sounds a similar dire warning for food security. Warming and acidifying oceans are threatening fish species on which many people depend for protein; and melting glaciers will result first in floods but eventually in loss of water supplies essential for agriculture (IPCC 2019b).

Climatic variability—such as the Little Ice Age that affected Europe from about the late fourteenth century to about 1860—has long been understood to exacerbate food shortages. During the coldest periods of the Little Ice Age, winters were frigid and prolonged, summers were unusually wet and chilly, and the growing season was shortened by several weeks. These conditions led to widespread crop failure, fungus outbreaks on grain, famine, war, riots, and epidemics (Pfister and Brázdil 2006). Throughout history, as Sheffield and Wood (2011) show, ancient empires and civilizations collapsed because climatic variabilities led to food shortages. Famous among them were the Maya civilization in Mexico, the Tang and Ming dynasties in China, and the Khmer Empire in Cambodia. Another important example of the collapse of an empire due in part to food shortages caused by climatic variability is the Roman Empire, as Kyle Harper describes in Chapter 8. He argues that the climatic conditions were not favorable for food production, further weakening the economic core of the empire. Food shortages led to famine in the city and in the surrounding food producing regions, which resulted in mass migrations into Rome that further weakened its ability to endure. Harper's specific historical example, which illuminates so clearly the impact of climate change on the fate of city and empire and on the disappearance of food sources and civilizations, suggests that we must pay very close attention to the already-severe impact of climate change and learn how to mitigate it if we do not want history to repeat itself at a larger scale.

Increased poverty

As global warming increases, poverty is expected to increase as well, especially among more vulnerable and disadvantaged populations. Given that poverty is closely linked with food insecurity, food insecurity will increase. Furthermore, the inundation of coastal lands that are currently being used for agriculture, as well as the loss of fish species from capture fisheries as the oceans warm, will add to higher food insecurity. Internal migration due to climate change is projected by the World Bank to increase to 143 million people in sub-Saharan Africa, Asia, and Latin America by 2050 (Rigaud et al. 2018), adding to the numbers of people who are no longer producing food but still desperately needing to eat. Conditions within refugee camps and slums on the outskirts of major cities in the non-Western world are already grim. They are only likely to worsen as increasing numbers of refugees as well as rural to urban migration continue to be common responses to food insecurity.

As we think about food insecurity in such settings, we must recognize that at the most local of scales, the individual and the family, it is associated with other insecurities, especially for women. As Adam Gilbertson argues in Chapter 2, food insecurity among slum dwellers in Kenya is not only the outcome of male partners preventing their wives from earning money, as may be the case in some traditional societies but, also, and perhaps more so, when men fail to provide for their families either by withholding money they have earned or not being part of the labor force at all. Gilbertson shows that in Bangla, Kenya, lack of money and food, regardless of the reasons, is intimately correlated to domestic violence and protracted abuse—sometimes of men but more often of women. Further, this chapter shows that money, food, gender, risk, and violence are closely connected to the power structure of the home and cycles of food insecurity. And when there is not enough money for food and hunger persists, the situation leads to intimate partner violence and at times to women running away from home, women and children are left in vulnerable positions.

Structural adjustment and austerity

Although many programs have been established to eradicate poverty, some of these efforts have actually exacerbated it and worsened the ability of people in developing and industrialized countries to access safe and nutritional foods. Especially pernicious effects on food insecurity can be traced to "structural adjustment programs" that were set up as conditions of loans from the International Monetary Fund (IMF) and World Bank (WB). The wealthier countries that controlled funds were able to impose harsh conditionalities on loan-seeking countries, ostensibly to bring them into the world market as full players, but in fact with devastating results, including opening up domestic markets to products from the wealthy countries. The conditions created food dependency by favoring markets for export crops over traditional agriculture that produced food for households, leaving people exposed to food shortages and price fluctuations. Conditionalities also dismantled grain reserves and prohibited further state investment in agriculture and most social services.

Much of the problem with agricultural restructuring began in the 1970s and 1980s as USAID promoted "agriculture for change," which subsidized non-traditional crops for export. In Costa Rica, for example, these subsidies weakened protections that had existed for farmers who produced for the domestic market, thus destabilizing both local markets and traditional ways of life. By the 2000s, many smallholder family farmers had been forced to sell their land to one of the few companies that had won large export contracts (Botella-Rodriguez 2018; Pietilainen and Otero 2019) and produced biofuels or food for the international market. Exports by these companies rose, and their profits soared. And while an increase in food exports resulted in an increase in real per capita GDP, the higher international food prices "induced a significant decrease in real per capita consumption and a significant increase in income inequality" (Arezki and Brückner 2011:11) resulting in both a "neoliberal diet" (Otero et al. 2015) and a widening gap between rich and poor in all countries subjected to structural adjustment contingencies (Amanor 2009; McMichael 2009; Moseley et al. 2010).

Effects of more recent austerity policies enacted under neoliberal regimes mimic the draconian impacts of structural adjustment. The objective of neoliberal policies is to liberalize or deregulate markets, under the belief that "free" markets result in greater good for all. However, as we see in most examples in this volume, the state, with the support of neoliberal regimes, intervenes regularly in food and agricultural markets; wealthier states get away scot-free while poorer states have severe penalties imposed on them.

In this volume, Charalampos Konstantinidis (Chapter 4), Murat Öztürk and colleagues (Chapter 5) and Birgit Schmook and colleagues (Chapter 6) describe the consequences of neoliberal policies in Greece, Turkey, and Nicaragua. In each case, these policies resulted in increased food insecurity. In the case of Greece (Chapter 4), especially after the 2007–08 crises, a series of structural adjustment programs have led to a dramatic increase in food prices and to the undermining of people's ability to govern their own food systems, effectively making food sovereignty in Greece impossible. Greece is particularly interesting because the structural adjustment programs that followed the food crises, and in which the government actively participated, followed in the footsteps of programs that aimed at changing Greece's long position on Europe's periphery. They clearly did not work very well before the 2007–08 crisis, and the new set of austerity plans and neoliberal policies that strengthened large players, centralized the food retail sector, and squeezed out small farmers further contributed to food insecurity. In Turkey (Chapter 5), too, many government policies compromised access to food. As with Greece, an examination of the troubled agricultural sector must also take a longer historical view. As the Turkish government decreased its subsidies to the agricultural sector in the beginning of 1980 and restricted employment in the public sector, small-scale farming was almost eliminated. In addition, increases in the price of transportation, storage, and waste—as well as neoliberal policies that weakened small- and medium-size traditional farms, forcing them to produce food for the international market—have led this once-thriving agricultural state to become food insecure. As Öztürk and colleagues show, this is particularly likely in countries such as Turkey, Iraq, and Syria, which have experienced military conflict, civil unrest, or war.

In Nicaragua (see Chapter 6), where lands have historically been in the hands of the very few, the rural population continues to be marginalized. Even when there was an attempt to redistribute lands and invest in agriculture during the Sandinista years of the 1980s, government policies favored modernization, at the expense of the already-marginalized landless peasants. As Schmook and colleagues show, food insecurity is a widespread problem in rural Nicaragua because much of the food provision in the countryside depends heavily on local agricultural production, and land distribution is so skewed. By orienting the economy toward export and disinvesting in small farms, Nicaragua's government has been responsible for the continued marginalization of farmers and for their food insecurity.

Another impact of neoliberal policies on food insecurity comes from transnational large-scale land acquisition, also known as "land-grabbing" or the new "global land rush" (Edelman et al. 2013). When governments, multinational corporations, and private firms either purchase or invest in large parcels of land in other countries for the purpose of agricultural production for world markets, small farmers are displaced, dispossessed, and disenfranchised. As the magnitude, scale, and pace of land-grabbing intensify, food production for the local market disappears, increasing the level of food insecurity and hunger among large populations (Wise 2019). But as Jampel Dell'Angelo and colleagues explain in Chapter 7, we cannot talk about land-grabbing without connecting it to water-grabbing, for they are intertwined. Water-grabbing, these authors argue, has even more severe impacts than losing access to land, especially in countries with high malnutrition that suffer from water scarcity. In many poorer countries the competition for water is harsh. Land investors can invest in hydraulic irrigation systems that small farmers could never afford; and, indeed, in the competition for water resources, even farmers who still own a small plot of land can never win. This dynamic invariably leads to dispossession of both water and lands. Therefore, land-grabbing and water-grabbing, which are inseparable from the corporate neoliberal agricultural model, have altered the agricultural employment sector, encouraged rural to urban migration, erased traditional rural food systems, and changed where food is sourced. The interplay of these effects has increased the disparity between people and regions, led to greater malnutrition in these areas, and resulted in social, economic, and political instability.

Responses to food insecurity

Food insecurity, whether at the local or national level, affects the well-being of the individual and the nation and demands a response at the state, community, and household levels. One of the primary determinants of a "failed state" is its inability to ensure food security and other social services (Newman 2009), yet almost all countries have pockets of food insecurity associated with poverty and marginalization. Responses begin with state-level policies, increasingly in coordination with the private sector (and in accordance with neoliberal influences) but include resistance and adaptation at the local level. When responses break down, the only alternative to starvation is flight. The form of response chosen by a government affects national identity and the ways that the population remembers periods of hunger, having deeper and more long-lasting impacts than only the alleviation of food insecurity, as we describe below.

Agribusinesses, and the governments and philanthropic organizations that have been captured by them, promise to ensure future food security sustainably for all. Yet they offer little beyond a continuation and intensification of the industrial agriculture that has resulted in massive land degradation, loss of biodiversity, over-exploitation and contamination of water supplies, and greenhouse-gas emissions (Wise 2019). A global debate about how best to assure food security is underway between the private sector and major international philanthropic organizations such as the Bill & Melinda Gates Foundation, and the swelling social movement for food sovereignty and agroecology (Patel 2008; Lawrence and McMichael 2012; Wise 2019). Policies adopted by the state determine which side is more likely to win this debate. At the international scale, the fight over policies plays out in the World Trade Organization and the Committee on World Food Security (CFS), the primary forum for international discussion and negotiation of issues affecting food security. Since reform of the CFS in 2009, civil society has equal voice with the private sector, representatives of UN agencies, scientific organizations, and international philanthropic organizations. Although national delegates are the only voting members of the CFS, they are able to hear arguments from all sides. Civil society brings to this debate the voices of frontline communities that suffer from food insecurity and violations of the right to food, allowing them to confront national delegates and others who purport to improve the prevalence of food security. Those voices report what has actually happened in their communities: rising hunger, loss of access to resources, criminalization of civil society and militarized responses by the state to peaceful protests. All guidelines and recommendations adopted by the CFS are mandated to be congruent with the right to food, which has been formally accepted through ratification of numerous international treaties and covenants by almost every country (CFS 2009).

Policies to increase food security

A distinctive set of policies that favor commercial interests ensues in neoliberal economies. The emphasis is on increasing production or productivity of crops and livestock, often through the most sophisticated science and technology. Thus, just as in the Green Revolution beginning in the 1940s, larger-scale farmers with more access to resources are most able to take advantage of these technologies, which leads to the widening gap between rich and poor. Neoliberalism also favors cash-cropping for export rather than household production for self-reliance, international trade of commodity crops, removal of regulations that were designed to protect labor and the environment, and privatization of services and land. In this policy model, the availability dimension of food security is emphasized, with little attention to accessibility, consumption, or stability dimensions. If adequate or arable land for increased production is not available within country borders, richer governments purchase or lease land from the poorest countries, usually in Africa or Latin America, where tenure rights are tenuous. For more than a decade, this has been the practice of the People's Republic of China (PRC) and Saudi Arabia. At other times, the national governments seek aid, sometimes enslaving their economies and mortgaging their futures as a result. In neoliberal economies the favored policy approach (indeed often the only policy approach) is market-based, assuming that producers, small businesses and consumers have adequate knowledge and capacity to make the right choices for themselves as well as for the future of the planet. These presumptions are seldom borne out in reality, however: without intervening public policies, wealth and power tend to accrue to the largest actors in the food system. As a result, gaps in food security deepen between wealthy and poor people. Depending on the region, the most powerful actors are transnational agribusinesses or retailers who can set prices that squeeze both producers and eaters (McIntyre et al. 2009).

In contrast to neoliberal policies, rights-based approaches emphasize improving access to food by providing jobs or direct payments so that people can afford to purchase it, increasing self-reliance rather than dependence on imports or expensive technology, and further enabling the democratic governance of resources and access to them by poor and marginalized people. In this volume, Dell'Angelo and colleagues (Chapter 7) discuss the need for better governance of natural resources in the context of land-grabbing and water-grabbing. They argue that assessments about food, water, and land are inseparable from rural justice, and "that issues of hunger and food security need to be addressed not only from the technical point of view but also [as] part of a dynamic that involves ethics, struggle, power dynamics, and politics" (p. 123, in this volume). At the international scale, the Voluntary Guidelines on the Responsible Governance of Tenure of Land, Fisheries, and Forests in the Context of National Food Security were designed and adopted by the CFS in 2012, after extensive negotiation among all parties. This agreement and other guidelines adopted by the CFS and a number of countries that use rights-based policies are examples of how food security can be improved dramatically through participatory actions that center on people who have not benefited from the neoliberal food system.

Policies to explicitly promote food sovereignty are in keeping with rights-based approaches (Kay et al. 2018). They arise from a commitment at the governmental level to democratic, transparent approaches that empower smallholders and offer a very different vision of the future of food systems than market-based policies provide. Policies supporting food sovereignty defend the rights of peasants, support the right to food and nutrition for all, and promote agroecology. Agroecology is most commonly conceptualized as a scientific discipline; a set of practices merging scientific, traditional, and indigenous knowledge; and a social movement (FAO 2019). Although governments that are more responsive to agribusiness than to small-scale farmers have resisted the rise of agroecology, it is widely seen now as an antidote to many of the environmental, social, and health problems associated with the industrial food system. FAO has organized several international and regional conferences on agroecology, established ten principles that apply to agroecological systems, and initiated a knowledge-sharing platform. Through its emphasis on soil health and crop diversity, agroecology has significant benefits for carbon sequestration, water-holding capacity, public health, and farmer livelihoods (Rosset and Altieri 2017). National research policy determines whether agroecology and food sovereignty will receive funding; with the growing influence of private corporations over research agencies in some countries, agroecology receives short shrift because it emphasizes reducing the use of purchased inputs in favor of biological and cultural pest controls, and applying compost or manure rather than purchased fertilizers. An increasing proportion of agricultural research is conducted by private corporations, or, as in the PRC's case, by privatized agencies that were formerly state-owned (Anderson 2018).

When disasters strike, governmental responses to food security are tested severely. Can the government provide food in emergencies and help food producers get back to "normal" food production practices, and will it merit the trust of its populations? Risk-communication policies are critical. In Chapter 12, Tomiko Yamaguchi and Joo-Young Jung discuss how one of the most serious instances of food contamination from nuclear fallout occurred in the months after the 2011 disaster in Fukushima. The Japanese government failed to adequately communicate about the dangers of food raised in radioactive soils and did little to alleviate fears in the following years, when little or no radiation was found in the environment. It failed to communicate effectively how rigorous the tests to detect radiation have been and how thorough the cleanup continues to be. If radiation is detected, produce is destroyed, and soils and water sources are thoroughly cleaned. As a result of poor government communication, people in the area, and at times as far as Tokyo, have been extremely reluctant to buy food produced in the once-lush agricultural prefecture. The Fukushima disaster and the perception of its aftermath have altered both food production and consumption patterns and left farmers in the region to struggle as their incomes have been depressed to below 2011 levels. While local residents now face a moral dilemma of whether to support their neighbor farmers or worry about food safety, Japan's government implemented new protocols that ensure securing foods in an aftermath of a disaster. However, *The Emergency Food Security Guidelines*, as Yamaguchi and Jung assert, still do not outline how to disseminate information about food security in times of a crisis of such magnitude. Unless it does so explicitly, the government's effort to provide access to safe and secure food supply will inevitably fall short (p. 208, in this volume).

In another case dealing with culturally influenced perceptions of food safety, Patricia Stapleton (Chapter 13) introduces the safety of genetically engineered organisms or GMOs (especially seeds) as a major barrier in setting up trade agreements between the United States and the EU. The United States, operating within the neoliberal model, insists that GMOs can boost food production and security, even though evidence for this is nonexistent, and increasing the quantity of food produced has very little impact on improving access by poor people. The EU is skeptical of GMOs for cultural, scientific, and political reasons and wary of allowing incursions into its territory. While EU policies are deemed protectionist by the US government, they stem from very different approaches to determining and managing risk: use of the precautionary principle versus a technocratic risk-management protocol. Not only are members of the EU not buying GMO products, but countries that export food to Europe fear that if they buy GMO products from the United States their crops may be contaminated and barred from entering European markets. This would affect the exporting countries' food producers and production. Thus, an exporting country may refuse food aid that could contaminate its seed supplies, even if it is facing food shortages. While it is not clear that a trade agreement to sell GM foods would affect food security within Europe, Stapleton suggests that such agreements could affect farmers in the United States and in EU countries, as well as in developing countries. Because the pesticides and fertilizers necessary in producing genetically modified foods, as well as the GM seed, are controlled by multinational corporations, free-trade agreements that open markets to GM food can put farmers at a more severe disadvantage, vis-à-vis agribusiness.

Other policies relevant to risk management for food producers and citizens, beyond those that apply to food safety, also have a powerful influence on food insecurity. Most industrialized countries have social-welfare safety nets for citizens, if not all residents, to help them either recover from individual, family, or community tragedies (ranging from illness to loss of employment) or overcome poverty. These safety nets vary depending on the state's prevailing ideology and financial resources. Under neoliberalism in the United States, UK, and Europe, food banking and other charitable organizations have grown dramatically, although they are ineffective in stemming growing food insecurity (Riches 2018). In contrast, under the Fome Zero initiative in Brazil, based on the right to food and nutrition, both moderate and severe food insecurity dropped to levels below that in the United States and the UK, as direct payments to poor people and programs to support food access and nutrition were implemented (Paes-Sousa and Vaitsman 2014; Cafiero et al. 2016; Chappell 2018).

At the same time as neoliberal states struggle to find ways to address poor access to food, the PRC has had a challenge of a different magnitude. With 20 percent of the world population—and with limited and unequally distributed arable land and water, as well as the experience of numerous instances of food shortages and famines in the modern period alone—China's responses to food insecurity have been manifold. Policies enacted have depended greatly on tradition and the ideological underpinnings of the regime at the time of famine. Pierre Fuller, in Chapter 9, analyzes the changing nature of the state's involvement in disaster relief, arguing that the party-state's responses since the establishment of the PRC in 1949 have been inconsistent with China's long past, when at times the emperor provided relief for the poor in the form of cash, coal, clothing, congee, and the transfer of grain from abundant regions to needy ones. At other times the ruling dynasty coordinated efforts to raise money for the hardest hit regions, controlled prices for grain, established distribution and refugee centers in regions with severe impacts, and sought international aid and charity relief. During the Great Leap Forward, however, alleviating hunger did not include government aid, although the Mao government did intervene. It stripped farming families of their legal rights to the grain that they produced and then directed it, through the newly established communes, to the urban industrial workforce. The new food production and distribution structures, which were initiated and supported by government policies, resulted in mass starvation. In the post-Mao years and to date, the government has returned to its earlier pattern of delivering foods and goods, but now mostly to the middle class. Such actions have been augmented by land grabs and land acquisition in Africa and Brazil (Hofman and Ho 2012; de L. T. Oliveira 2017), "development outsourcing" (Hofman and Ho 2012), excessive use of fertilizers, and charitable donations mobilized by the growing civil society. The current need to feed more than a billion and a half people affects not only those who live in the PRC but others around the globe as well.

Popular resistance and adaptation

Farmers and rural communities worldwide are finding it more difficult to manage risks associated with food production. They face climate change with its droughts and floods, land and water grabs, seed patents, low prices for their goods, and often political instability. At the same time, their social and economic support systems have weakened, providing farming communities with limited ways to overcome their poverty (Pelletier et al. 2016) and avoid "hunger seasons" (Béné et al. 2016:123). As we see throughout this volume, food insecurity is present in different settings and historical contexts and manifested differently depending on time and place. While UN agencies such as FAO and IFAD seek ways to help rural communities build the capacity to cope with and adapt to a changing world, all who are food insecure must find creative ways to adapt and alter their predicament. We can see such responses in almost every chapter in this volume, ranging from household and farm to national and global. At perhaps the most micro scale of responses considered here, Adam Gilbertson (Chapter 2) depicts how women exert power and secure food for themselves and their children in unequal domestic relationships. Jessie Mazar and Teresa Mares (Chapter 3) show how labor migrants on the Vermont–Canada border, especially those whose mobility is limited for fear of being deported and therefore cannot easily access traditional foods, are able to have such foods when they plant their own gardens on the farms at which they work. These gardens, Mazar and Mares conclude, are the source of both home-food and spaces of autonomy amidst a landscape of fear (p. 45, in this volume). In Greece (Chapter 4), Konstantinidis tells us about attempts of a number of movements and grassroot initiatives at rebuilding the agricultural sector by reconnecting farmers and consumers and challenging both existing food production and distribution in the crisis-ridden country. Sadly, however, many of these efforts have been crushed by neoliberal policies.

In China and France resistance and adaptation are marked by remembering times of food shortages and, in France at least, also by how people actively resisted the shortages. In China, Ellen Oxfeld (Chapter 10) shows how the memory of times of food shortage is crafted into a narrative of survival against the odds, giving people a measure of dignity even under severe oppression by their own government or by an invading force. She shows how villagers' food and food practices, especially during holidays, embody historical memory of times of great hardship and at the same time mark an historic transformation from food shortages to food abundance. Paula Schwartz (Chapter 11) depicts ways that French citizens used humor to soften the pain of starvation under German occupation. She details the adjustments French urbanites made during the Second World War, when food was so scarce. They changed their modes of cooking, took to raising their own chickens and rabbits on their balconies, and adapted to a war-time diet. At the same time, however, as food became less available and the lines in front of grocery stores got longer, these queues became volatile spaces, inhabited mostly by women, and sites of food protests (p. 191, in this volume). As both the Oxfeld and Schwartz chapters show, even in times of severe food insecurity and even when options are sorely limited, people derive meaning in their lives and exert agency.

A common reaction or response to food insecurity is active resistance to the conditions that create it, whether these are actions of politicians or pressure by other countries. At a local level, high food prices during the 2007–08 food price crisis sent hungry people to the streets in protest in the Philippines, the Caribbean, the Middle East, many parts of Asia, and more than twenty-five countries in Africa (McMichael 2009; Bush 2010; Berazneva and Lee 2013; Smith 2014; Wischnath and Buhaug 2014), threatening the political and social status quo of both the local and national levels. In Haiti, for example, the government collapsed (Bohstedt 2014); in Guatemala violence increased (Pietilainen and Otero 2019); and in the Middle East, the Arab Spring was launched in response to the sharp rise of food insecurity in 2008 (Zarayk 2011). Protests were not only about the price of food, although this was the impetus, but also about long-established racial and gender injustices, class inequality, political repression, unequal economic development, and allocation of resources (Uraguchi 2010; Gana 2012). Lack of food was the match to the tinder of injustice, igniting them into full-blown rebellion.

Short of full-fledged rebellions and riots in the streets and demanding that the state make changes to increase food access, there have been other ways by which to ensure access to food. In the case of Greece after the global financial crisis of 2007–08, people turned inward to their neighbors and communities for help. Greeks, explains Konstantinidis in Chapter 4, sought to reshape their agriculture and the food system by creating a sharing economy at the community scale; there were efforts, he writes, to connect "small producers and consumers, circumventing middlemen and supermarkets" (p. 73, in this volume) and engaging in the global food sovereignty movement. Similarly, and at a more local level, Mazar and Mares show in Chapter 3 how Mexican migrant farmworkers on the US northern borderlands, who have not been able to access culturally appropriate food, have responded with resilience and creativity. They too have created spaces where food sovereignty and sharing exist by planting vegetable gardens with seeds from home, thus bringing "home" and food to their new space.

When we look at responses with a long historical lens, as Harper does in Chapter 8, we can assume that in periods of colder, wetter weather less favorable to food production, farmers had to adapt. Harper thus explores the relationship between responses to food insecurity and the fall of Rome. At the same time that farmers responded to their immediate situation, the empire did as well, especially through shipping food from one part (Egypt) to another (Constantinople). The responses of the time, however, were short-lived, allowing perhaps only a few additional centuries before Rome fell. Food shortages were a contributor.

Food insecurity and national identity

The problem of food insecurity goes beyond the inability to feed people. It also involves national identity, where abundance is a reason for national pride and where insecurity (i.e., dependence) may mean embarrassment and a sense of defeat. Food is important beyond sustenance; it plays a pivotal role in the construction of national and cultural identities, for it is "packed with social, cultural, and symbolic meaning" (Bell and Valentine 1997:3). Food is a vehicle through which individuals participate in and embody their national and cultural allegiances, and a way through which the state prescribes nationalist meaning to the staples that are grown and sold. Therefore, in times of food insecurity, when the state becomes dependent on others, it may react by imposing protectionist food policies, as Russia and Indonesia have done. After 1991, in its deliberate efforts of nation building, Russia was determined to construct a deeply rooted nationalist food culture and rhetoric (Caldwell 2002); and, as in the Soviet era, it now saw buying imported food as unpatriotic. Therefore, it engaged in a "Buy Russia" campaign, distinguishing foods as "Ours" and "Not Ours" (Caldwell 2002), and banned food imports from countries that had placed sanctions on it. Food that slipped through the cracks was destroyed. In particular, the government banned luxury foods to which the new middle class had become accustomed. The ban, and the food policies in general, coincided with massive subsidies for the production and export of grains (Caldwell 2002; Wegren 2011) and were rooted in nationalism and renewed efforts at nation building.

In Indonesia, too, food self-sufficiency has been a tool for strengthening nationalism. For years much of the country's political rhetoric has focused on food security and on achieving national self-sufficiency. The 2007–08 crisis and the increase in world food prices, as well as its dependence on foreign producers, provided the government the political impetus it needed to engage in protectionist policies and in a "new nationalism" that focused almost exclusively on increased food production. The new nationalism was a "defensive response to deepening food insecurity within Indonesia and to its poor economic and social developments relative to neighboring countries" (Neilson and Wright 2017: 136). Even though the aim was to bring food sovereignty, the Indonesian government never defined what this entailed and paid little to no attention to peasants and their rights. In other words, food sovereignty (or the goal of self-sufficiency for farmers) was "a rhetorical device to strengthen the role and function of the state" (Neilson and Wright 2017: 138).

When access to food is compromised, political instability may ensue, particularly in low-income countries (Arezki and Brückner 2011; Berazneva and Lee 2013). As domestic agricultural systems were decimated by structural adjustment programs, poorer countries became more reliant on food imports, leading to greater nationalist rhetoric, protectionist policies, and hoarding of food by countries, such as the Philippines, which had previously been major exporters of staples. Consequences were dire for countries that had become newly dependent on imported food. But the connection between national identity and food also involves memories of times when people went hungry during disasters—whether precipitated by "natural" events or not. These memories remain part of the culture and are imprinted onto the national, familial, and individual DNA. As Ralph Thaxton (2008) suggests, such memories are "embodied" (see Chapters 7 and 8 in this volume).

The power of memory

Most studies on food and identity illuminate how food and power are inseparable; how power separates those who have access to food from those who do not (Mintz 1996; Holtzman 2006; Anderson and Leach 2019) and how, for many, food can symbolize dependence, domination, hierarchy, oppression, and identity politics broadly defined (Mintz 1994; Weinreb 2017; Avieli 2018). As important as food has been in the construction of national identity (Anderson 1983 and 1991; Hobsbawm 1983; Appadurai 1988, Weinreb 2017), the lack of food has shaped nations as well. If it is not just the actual food shortages and their impact on health and growth that shape the nation and its parts, then it is certainly the memory of hunger and lack of access to food. Regardless of whether famines were the result of droughts, war, or the use of food as a weapon, the memory of them shaped Americans during the Dust Bowl, sub-Saharan Africans during the Sahelian droughts, the Chinese during four separate recent famines, Europeans during the Second World War; Palestinians in Gaza under siege; and Africans, Syrians, Iraqis, and others in the latest iterations of armed conflicts and climate refugees.

Oxfeld in Chapter 10 and Schwartz in Chapter 11 recount memories of food scarcity in a Chinese village during the collective era and the Great Leap famine, and in France during the Second World War. Food scarcity left its mark on the war generation in France and the famine generation in the PRC and has affected the next generation as well. Oxfeld shows that food practices in the particular Chinese village she describes not only embody historical memory but have become markers of the transformation that the PRC underwent from being food-short to food-abundant. She shows how villagers remember the past through recalling memories of specific foods—their taste, smell, and texture. They remember foods being more bitter and less smooth because they were cooked without ingredients, like oil, that were not available during the famine. Schwartz shows how stories of hunger continued to be told at French family gatherings and were burned into the consciousness of those who grew up in the 1950s and beyond, so much so that staples like the Jerusalem artichoke and rutabaga were avoided at all costs, and people continued to exhibit food-hoarding tendencies well past the time of shortages. In similar ways, migrants may long for foods from their home cultures and plant gardens of familiar vegetables to give their households the meals that remind them of home, as do the Mexican migrants in the United States, described by Mazar and Mares in Chapter 3.

Thus responses to food insecurity range from the national to the deeply personal level, and impacts may endure well beyond a hunger season, for multiple generations. The type and effectiveness of response can restore health, help to build community goodwill, and create a strong sense of national identity or lead to social upheaval and the overthrow of a government.

A final conclusion apparent in the compilation of chapters in this volume is that scholarship and analysis are themselves ways to respond to food insecurity by increasing awareness of its causative factors, building an understanding of the range of impacts over generations, and critically examining proposed solutions. The complexity of causation, impacts, and viable solutions requires scholarship from many disciplines or even outside disciplines: transdisciplinary scholarship that bridges academics and social movements, indigenous peoples, and non-governmental associations is enhancing food studies by breaking through assumptions about how people can live and work together, and how they can sustainably co-manage the resources needed for food production (IPES-Food 2015). In particular, knowledge of agroecology and food sovereignty, as counterpoints to the scourge of industrial agriculture, neoliberal policies, and accompanying exploitation of land, resources, and labor, has developed rapidly over the last couple of decades, with more academic scholars working with the people who are living these alternatives. "Engaged political economy" in which scholars work with marginalized or vulnerable communities to examine the power dynamics that lead to their marginalization has been an especially promising approach (Anderson and Leach 2019). Investigation of power in food systems is benefiting from feminist perspectives, analysis of different forms of knowledge, power mapping, critical digital studies, living labs, diverse economies, and studies of the co-production of knowledge and nature as well (Duncan et al. 2019).

While population growth, climate change, vanishing biodiversity, and war are rapidly escalating the threats of food insecurity, alternatives that would allow all people to have access to sufficient healthy food are within grasp. Like many of the hopeful signs on the horizon, these depend on mustering a huge amount of political will in a very short time period: difficult, but not impossible. We hope the chapters in this volume will help to illuminate this path.

References

Alkon, Alison Hope and Julian Agyeman (Eds.). 2011. *Cultivating Food Justice: Race, Class and Sustainability*. Cambridge, MA: MIT Press.

Amanor, Kojo. 2009. "Global food chains, African smallholders and World Bank governance." *Journal of Agrarian Change* 9(2): 247–262.

Anderson, Benedict. 1983, 1991. *Imagined Communities*. London: Verso.

Anderson, Molly D. 2018. "Scientific knowledge of food and agriculture in public institutions: Movement from public to private goods," pp. 185–202. In Jose Luis Vivero Pol, Tomaso Ferrando, Olivier de Schutter, and Ugo Mattei (Eds.). *Routledge Handbook of Food as a Commons: An Alternative Narrative and Political Praxis to Cultivate and Eat*. London: Routledge.

Anderson, Molly D., with Bama Athreya. 2015. "Improving the well-being of food system workers," pp. 108–127. In *Advancing Health and Well-being in Food Systems Compendium: Strategic Opportunities for Funders*. Global Alliance for the Future of Food.

Anderson, Molly D., and Melissa Leach. 2019. "Transforming food systems: The potential of engaged political economy." *Institute of Development Studies Bulletin* 50(2): 131–146.

Anderson, Molly, Nicholas Nisbett, Chantal Clément, and Jody Harris. 2019. "Introduction: Valuing different perspectives on power in the food system." *IDS Bulletin* 50(2): 1–12.

Appadurai, Arjun. 1988. "How to make a national cuisine: Cookbooks in contemporary India." *Comparative Studies of Society and History* 30(1): 3–24.

Arezki, Rabha, and Markus Brückner. 2011. *Food Prices and Political Instability*. IMF Working Paper.

Avieli, Nir. 2018. *Food and Power: A Culinary Ethnography of Israel*. Oakland: University of California Press.

Baker, Lauren, Barbara Gemmill-Herren, and Fabio Leipper. 2019. *Beacons of Hope: Accelerating Transformations to Sustainable Food Systems*. Global Alliance for the Future of Food and Biovision Foundation for Ecological Development. https://foodsystemstransformations.org/wp-content/uploads/2019/08/BeaconsOfHope_Report_082019.pdf.

Bell, David, and Gill Valentine. 1997. *Consuming Geographies: We Are Where We Eat*. London: Routledge.

Béné, Christopher, Derek Headey, Lawrence Haddad, and Klaus von Grebmer. 2016. "Is resilience a useful concept in the context of food security and nutrition programmes? Some conceptual and practical considerations." *Food Security* 8: 123–138.

Berazneva, Julia, and David Lee. 2013. "Explaining the African food riots of 2007–2008: An empirical analysis." *Food Policy* 39: 28–39.

Bohstedt, John. 2014. "Food Riots and the Politics of Provision in World History." *Working Papers* No. 444, Brighton, UK: Institute of Development Studies.

Bonanno, Alessandro and Steven Wolf. 2018. *Resistance to Neoliberal Agri-Food: A Critical Analysis*. New York: Routledge.

Botella-Rodriguez, Elisa. 2018. "Costa Rica's outward-looking development: From 'Agriculture of Change' to food insecurity." *Bulletin of Latin American Research* 1: 57–72.

Bush, Ray. 2010. "Food riots: Poverty, power and protest." *Journal of Agrarian Change* 10(1): 119–129.

Cafiero, Carlo, Mark Nord, Sara Viviani, Mauro Eduardo Del Grossi, Terri Ballard, Anne Kepple, Meghan Miller, and Chiamaka Nwosu. 2016. *Methods for Estimating Comparable Prevalence Rates of Food Insecurity Experienced by Adults Throughout the World.* Rome: FAO.

Caldwell, Melissa. 2002. "The taste of nationalism: Food politics in post-socialist Moscow." *Ethnos* 67(3): 295–319.

Carolan, Michael S. 2018. *The Food Sharing Revolution: How Start-ups, Pop-ups, and Co-ops Are Changing the Way We Eat.* Washington, DC: Island Press.

Chappell, M. Jahi. 2018. *Beginning to End Hunger: Food and the Environment in Belo Horizonte, Brazil, and Beyond.* Oakland: University of California Press.

Clapp, Jennifer. 2014. "Food security and food sovereignty: Getting past the binary." *Dialogues in Human Geography* 4(2): 206–211.

Clapp, Jennifer and S. Ryan Isakson. 2018. *Speculative Harvests: Financialization, Food, and Agriculture.* Black Point, Nova Scotia: Fernwood Press.

Committee on World Food Security (CFS). 2009. *Reform of the Committee on World Food Security. CFS: 2009/2.* Rome: FAO.

de L. T. Oliveira, Gustavo. 2018. "Chinese land grabs in Brazil? Sinophobia and foreign investments in Brazilian soybean agribusiness." *Globalizations* 15(1): 114–133.

Dekeyser, Koen, Lise Korsten, and Lorenzo Fioramonti. 2018. "Food sovereignty: Shifting debates on democratic food governance." *Food Security* 10: 223–233.

Desmarais, Annette. 2007. *La Vía Campesina: Globalization and the Power of Peasants.* Halifax: Fernwood.

Dieterle, Michael. 2015. "Framing Food Justice," pp. 3–20. In J. Michael Dieterle (ed.). *Just Food: Philosophy, Justice and Food.* London & New York: Rowman and Littlefield.

Duncan, Jessica, Charles Z. Levkoe, and Ana Moragues-Faus. 2019. "Envisioning new horizons for the political economy of sustainable food system." *IDS Bulletin* 50(2): 37–56.

DuPuis, Melanie, Jill Lindsey Harrison, and David Goodman. 2013. "Just food?" pp. 283–308. In Alison Hope Alkon and Julian Agyeman (eds.). *Cultivating Food Justice: Race, Class and Sustainability*. Cambridge, MA: MIT Press.

Edelman, Marc. 2014a. "Food sovereignty: Forgotten genealogies and future regulatory challenges." *Journal of Peasant Studies* 41(6): 959–978.

Edelman, Marc. 2014b. "The next stage of the food sovereignty debate." *Dialogues in Human Geography* 4(2): 182–184.

Edelman, Marc, Carlos Oya, and Saturnino M. Borras, Jr. 2013. "Global land grabs: Historical processes, theoretical and methodological implications and current trajectories." *Third World Quarterly* 34(9): 1517–1531.

Edelman, Marc, Tony Weis, Amita Baviskar, Saturnino M. Borras Jr, Eric Holt-Giménez, Deniz Kandiyoti, and Wendy Wolford. 2014. "Introduction: Critical perspectives on food sovereignty." *Journal of Peasant Studies* 41(6): 911–931.

Environmental Justice Foundation. 2019. *Blood and Water: Human Rights Abuse in the Global Seafood Industry*. London.

FAO. 2019. *Agroecology Knowledge Hub*. www.fao.org/agroecology/en/.

FAO, IFAD, UNICEF, WFP, and WHO. 2019. *The State of Food Security and Nutrition in the World 2019. Safeguarding Against Economic Slowdowns and Downturns*. Rome: FAO.

Food Chain Workers Alliance and Solidarity Research Cooperative. November 2016. *No Piece of the Pie: U.S. Food Workers in 2016*. Los Angeles: Food Chain Workers Alliance.

Gana, Alia. 2012. "The rural and agricultural roots of the Tunisian Revolution: When food security matters." *International Journal of Sociology of Agriculture and Food* 19(2): 201–213.

Genoways, Ted. 2014. *The Chain: Farm, Factory and the Fate of Our Food*. New York: Harper.

Gilbert, Jess, Gwen Sharp, and M. Sindy Felin. 2002. "The loss and persistence of Black-owned farms and farmland: A review of the research literature and its implications." *Southern Rural Sociology* 18(2): 1–30.

Herman, Agatha, and Mike Goodman. 2018. "New spaces of food justice." *Local Environment* 23(11): 1041–1046.

Hobsbawm, Eric. 1983. "Introduction: Inventing traditions," pp. 1–14. In Eric Hobsbawm, T Ranger (eds.). *The Invention of Tradition*. Cambridge: Cambridge University Press.

Hofman, Irna, and Peter Ho. 2012. "China's 'Developmental Outsourcing': A critical examination of Chinese global 'land grabs' discourse." *The Journal of Peasant Studies* 39(1): 1–48.

Holt-Giménez, Eric. 2017. *The Foodie's Guide to Capitalism: Understanding the Political Economy of What We Eat*. New York: Monthly Review Press.

Holt-Giménez, Eric, and Yi Wang. 2011. "Reform or transformation? The pivotal role of food justice in the U.S. food movement." *Race/Ethnicity: Multidisciplinary Global Contexts* 5(1).

Holtzman, Jon. 2006. "Food and memory." *Annual Review of Anthropology* 35(1): 361–378.

IPCC. 2019a. "Climate change and land: An IPCC special report on climate change, desertification, land degradation, sustainable land management, food security, and greenhouse-gas fluxes in terrestrial ecosystems." www.ipcc.ch/report/srccl/

IPCC. 2019b. "The ocean and cryosphere in a changing climate." www.ipcc.ch/report/srocc/

IPES-Food. 2015. "The new science of sustainable food systems: Overcoming barriers to food systems reform." International Panel of Experts on Sustainable Food Systems. www.ipes-food.org/_img/upload/files/NewScienceofSusFood.pdf

IPES-Food. 2017. "Unravelling the food-health nexus: Addressing practices, political economy and power relations to build healthier food systems." www.ipes-food.org/_img/upload/files/Health_FullReport(1).pdf

Jarocz, Lucy. 2014. "Comparing food security and food sovereignty discourses." *Dialogues in Human Geography* 4(2): 168–141.

Kay, Sylvia, Emily Mattheisen, Nora McKeon, Paola De Meo, and Ana Moragues-Faus. 2018. "Public policies for food sovereignty." Food for Thought No. 1, Amsterdam: Transnational Institute.

Landeker, Hannah. 2011. "Food as exposure: Nutritional epigenetics and the new metabolism." *BioSocieties* 6(2): 167–194.

Lappé, Frances Moore, Jennifer Clapp, Molly Anderson, Robin Broad, Ellen Messer, Thomas Pogge, and Timothy Wise. 2013. "How we count hunger matters." *Ethics and International Affairs* 27(3):251–259.

Lappé, Frances Moore, and Joseph Collins. 2015. *World Hunger: 10 Myths*. New York: Grove Press.

Lawrence, Geoffrey, and Philip McMichael. 2012. "The question of food security." *International Journal of the Sociology of Agriculture and Food* 19(2): 135–142.

Levkoe, Charles Z., Josh Brem-Wilson, and Colin R. Anderson. 2018. "People, power, change: three pillars of a food sovereignty research praxis." *Journal of Peasant Studies*. Published online September 26, 2018.

Masson-Delmotte, Valérie, Panmao Zhai, Hans-Otto Pörtner, Debra Roberts, James Skea, Priyadarshi Shukla, Anna Pirani, Wilfran Moufouma-Okia, Clotilde Péan, Roz Pidcock, Sarah Connors, Robin Matthews, Yang Chen, Xiao Zhou, Melissa Gomis, Elizabeth Lonnoy, Thomas Maycock, Melinda Tignor, and Tim Waterfield (eds.). 2018. "Global warming of 1.5°C. An IPCC Special Report on the impacts of global warming of 1.5°C above pre-industrial levels and related global greenhouse-gas emission pathways, in the context of strengthening the global response to the threat of climate change, sustainable development, and efforts to eradicate poverty." Geneva: World Meteorological Organization.

McIntyre, Beverly, Hans Herren, Judi Wukhungu, and Robert Watson. 2009. *Agriculture at a Crossroads*. Global Report of the International Assessment of Agricultural Knowledge, Science & Technology for Development. Washington, DC: Island Press.

McMichael, Philip. 2009. "A food regime analysis of the 'world food crisis.'" *Agriculture and Human Values* 26: 281–295.

McMillan, Tracie. 2012. *The American Way of Eating: Undercover at Walmart, Applebee's, Farm Field and the Dinner Table*. New York: Scribner.

Mintz, Sidney. 1994. "Eating and being: What food means," pp. 102–115. In Barbara Harris-White and Sir Raymond Hoffenberg (eds.). *Food: Multidisciplinary Perspectives*. Oxford: Blackwell.

Mintz, Sidney. 1996. *Tasting Food, Tasting Freedom: Excursions into Eating, Culture, and the Past.* Boston: Beacon Press.

Moragues-Faus, Ana. 2017. "Problematising justice definitions in public food security debates: Towards global and participative food justices." *Geoforum* 84: 95–106.

Moseley, William, Carney Judith, and Laurence Becker. 2010. "Neoliberal policy, rural livelihoods, and urban food security in West Africa: A comparative study of The Gambia, Côte d'Ivoire, and Mali." *Proceedings of the National Academy of Sciences* 107(13): 5774–5779.

Myers, Samuel S., Antonella Zanobetti, Itai Kloog, Peter Huybers, Andrew D. B. Leakey, Arnold J. Bloom, Eli Carlisle, Lee H. Dietterich, Glenn Fitzgerald, Toshihiro Hasegawa, N. Michele Holbrook, Randall L. Nelson, Michael J. Ottman,Victor Raboy, Hidemitsu Sakai, Karla A. Sartor, Joel Schwartz, Saman Seneweera, Michael Tausz, and Yasuhiro Usui. 2014. "Increasing CO_2 threatens human nutrition." *Nature* 510: 139–142.

Neilson, Jeff, and Josephine Wright. 2017. "The state and food security discourses of Indonesia: Feeding the *bangsa*." *Geographical Research* 55(2): 131–143.

Nestle, Marion. 2002. *Food Politics: How the Food Industry Influences Nutrition & Health.* Oakland: University of California Press.

Newkirk, Vann R. II. 2019. "The great land robbery: The shameful story of how one million black families have been ripped from their farms." *The Atlantic* (September).

Newman, Edward. 2009. "Failed states and international order: Constructing a post-Westphalian world." *Contemporary Security Policy* 30(3): 421–443.

Nyéléni Forum. 2007. *Declaration of Nyéléni* https://nyeleni.org/spip.php?article290, accessed March 22, 2019.

Otero, Gerardo, Gabriela Pechlaner, Giselle Liberman, and Efe Can Gürcan. 2015. "Food security and inequality: Measuring the risk of exposure to the neoliberal diet." *Simons Papers in Security and Development* No. 42/2015.

Paes-Sousa, Romulo, and Jeni Vaitsman 2014. "The Zero Hunger and Brazil without Extreme Poverty programs: A step forward in Brazilian social protection policy." *Ciência & Saúde Coletiva* 19(11): 4351–4360.

Patel, Raj. 2008. *Stuffed and Starved: Markets, Power and the Hidden Battle for the World Food System*. London: Portobello Books Ltd.

Pelletier, Bernard, Gordon Hickey, Kimberli Bothi, and Andrew Mude. 2016. "Linking rural livelihood resilience and food security: An international challenge." *Food Security* 8: 469–476.

People's Knowledge Editorial Collective. 2016. *People's Knowledge and Participatory Action Research: Escaping the White-walled Labyrinth*. Practical Action Publishing.

Pfister, Christian, and Rudolf Brázdil. 2006. "Social vulnerability to climate in the 'Little Ice Age': An example from Central Europe in the early 1770s." *Climate of the Past, European Geosciences Union* 2(2): 115–129.

Pietilainen, Emma, and Gerardo Otero. 2019. "Power and dispossession in the neoliberal food regime: Oil palm expansion in Guatemala." *Journal of Peasant Studies,* 46: 1142–1166.

Pimbert, Michel P. 2006. "Transforming knowledge and ways of knowing for food sovereignty." London: International Institute for Environment and Development.

Pimbert, Michel P., Boukary Barry, Anne Berson, and Khanh Tran-Thanh. "Democratising agricultural research for food sovereignty in West Africa." 2010. Bamako and London: IIED, CNOP, Centre Djoliba, IRPAD, Kine Conseils, URPAD.

Popkin, Barry M. 1993. "Nutritional patterns and transitions." *Population and Development Review* 19(1): 138–157.

Rawls, John. 1971. *A Theory of Justice*. Cambridge, MA: Harvard University Press.

Riches, Graham. 2018. *Food Bank Nations: Poverty, Corporate Charity and the Right to Food*. London: Routledge.

Rigaud, Kanta Kumari, Alex de Sherbinin, Bryan Jones, Jonas Bergmann, Viviane Clement, Kayly Ober, Jacob Schewe, Susana Adamo, Brent McCusker, Silke Heuser, and Amelia Midgley. 2018. *Groundswell: Preparing for Internal Climate Migration*. Washington, DC: World Bank.

Rosset, Peter. 2006. *Food Is Different: Why We Must Get the WTO Out of Agriculture*. London: Zed Books.

Rosset, Peter M., and Miguel A. Altieri. 2017. *Agroecology: Science and Politics*. Black Point, Nova Scotia: Fernwood Publishing.

Sage, Colin. 2014. "Food security, food sovereignty and the special rapporteur: Shaping food policy discourse through realising the right to food." *Dialogues in Human Geography* 4(2): 195–199.

Sandel, Michael. 1984. "The procedural republic and the unencumbered self." *Political Theory* 12(1): 81–96.

Sbicca, Joshua. 2012. "Growing food justice by planting an anti-oppression foundation: Opportunities and obstacles for a budding social movement." *Agriculture, Food and Human Values* 29(4): 455–466.

Sbicca, Joshua. 2018. *Food Justice Now!* Minneapolis: University of Minnesota Press.

Schlosser, Eric. 2008. *Fast Food Nation. The Dark Side of the All-American Meal.* Boston: Houghton Mifflin Harcourt.

Scoville, J. Michael. 2015. "Framing food justice," pp. 3–20. In J. M. Dieterle (ed). *Just Food: Philosophy, Justice & Food.* London & New York: Rowman & Littlefield.

Sen, Amartya. 1999. *Development as Freedom.* Oxford: Oxford University Press.

Sheffield, Justin, and Eric Wood. 2011. *Drought: Past Problems and Future Scenarios*, London: Routledge, Earthscan.

Smith, Todd. 2014. "Feeding unrest: Disentangling the causal relationship between food price shocks and sociopolitical conflict in urban Africa." *Journal of Peace Research* 51(6): 679–695.

Thaxton, Ralph. 2008. *Catastrophe and Contention in Rural China: Mao's Great Leap Forward Famine and the Origins of Righteous Resistance in Da Fo Village.* Cambridge: Cambridge University Press.

Uraguchi, Zenebe Bashaw. 2010. "Food price hikes, food security, and gender equality: Assessing the roles and vulnerability of women in households of Bangladesh and Ethiopia." *Gender and Development* 18(3): 491–501.

Vivero-Pol, Jose Luis. 2019. "The idea of food as a commons: Multiple understandings for multiple dimensions of food," pp. 25–41. In Jose Luis Vivero Pol, Tomaso Ferrando, Olivier de Schutter, and Ugo Mattei (eds.). *Routledge Handbook of Food as a Commons: An Alternative Narrative and Political Praxis to Cultivate and Eat.* London: Routledge.

Walzer, Michael. 1990. "The communitarian critique of liberalism." *Political Theory* 18(1): 6–23.

Wegren, Stephen. 2011. "Food security and Russia's drought." *Eurasian Geography of Economics* 51(1): 140–156.

Weidman, Dennis. 2012. "Native American embodiment of the chronicities of modernity: Reservation food, diabetes, and the metabolic syndrome among the Kiowa, Comanche, and Apache." *Medical Anthropology Quarterly* 26(4): 595–612.

Weinreb, Alice. 2017. *Modern Hungers: Food and Power in the Twentieth-century Germany.* New York: Oxford University Press.

Wischnath, Gerdis, and Halvard Buhaug. 2014. "Rice or riots: On food production and conflict severity across India." *Political Geography* 43(11): 6–15.

Wise, Tim. 2019. *Eating Tomorrow: Agribusiness, Family Farmers and the Battle for the Future of Food.* New York: The New Press.

Wise, Tim, and Sophia Murphy. 2013. "Resolving the food crisis: assessing global food policy reforms since 2007." Institute for Agriculture and Trade Policy (IATP) and Global Development and Environment Institute (GDAE).

Zarayk, Rami. 2011. *Food, Farming and Freedom: Sowing the Arab Spring.* Washington, DC: Just World Books.

Zhao, Chuang, Bing Liu, Shilong Piao, Xuhui Wang, David B. Lobell, Yao Huang, Mengtian Huang, Yitong Yao, Simona Bassu, Philippe Ciais, Jean-Louis Durand, Joshua Elliott, Frank Ewert, Ivan A. Janssens, Tao Li, Erda Lin, Qiang Liu, Pierre Martre, Christoph Müller, Shushi Peng, Josep Peñuelas, Alex C. Ruane, Daniel Wallach, Tao Wang, Donghai Wu, Zhuo Liu, Yan Zhu, Zaichun Zhu, and Senthold Asseng. 2017. "Temperature increase reduces global yields of major crops in four independent estimates." *Proceedings of the National Academy of Science* 35: 9326–9331.

2 Causes and consequences of *njaa* (hunger) in the household

Food security and intimate partner violence within an informal settlement in Mombasa, Kenya

Adam Gilbertson

Introduction

One out of every eight people living today resides in a "slum" (United Nations Human Settlements Programme 2016). In low- and middle-income countries, at least 30 percent of urban residents live in communities judged to be unfit for healthy life; in sub-Saharan Africa, informal settlements (known as "townships" in South Africa and "slums" in Kenya) are home to over half of all urban residents (United Nations Human Settlements Programme 2016). Informal settlements are interstitially or peripherally located, marginalized urban spaces characterized by poverty, unemployment, lack of infrastructure, crowded living, inadequate housing, tenuous land rights, infectious diseases, and intimate partner violence (IPV) (Swart 2012). Frequently, these communities are located beside or in between elite residential areas—the former's residents often work for the latter's as cooks, gardeners, "house help," and as manual laborers. Unlike refugee camps or those built for internally displaced persons (IDPs), informal settlements in Kenya function as permanent communities, often with extensive histories dating back decades to colonial times and policies of racial segregation (K'Akumu and Olima 2007). In urban Kenya, and throughout the global south, informal settlements contain households with the greatest degree of food insecurity (Crush and Frayne 2010; Crush and Frayne 2011; Faye et al. 2011; Kimani-Murage et al. 2014). Although the degree to which a household (or an individual) is food secure is always relative, for the majority of residents, it is a daily struggle (often an impossible one) to achieve safe, reliable, sustainable access to sufficient, affordable, and preferred foodstuffs and the resources to prepare and consume them (e.g., fuel, water, cooking oil, and cookware).

The many drivers of food insecurity within informal settlements include all those one would expect to find in environments plagued by poverty, unemployment, social and political marginalization, pollution, inadequate education, corruption, and a perennially struggling economy. In the simplest terms, food insecurity in an informal settlement results from the lack of sufficient cash in hand to make daily food purchases at local markets (refrigeration is a rare luxury). Yet, the reasons why entire households (or certain members) may fail to access sufficient food are often more complicated. Individual access to food and other resources within household units tends to depend on everyday interpersonal interactions and exchanges between household members, which often result in the unequal division of resources. The distribution of food, or money to buy food, may depend on factors such as gender, age, and perceived economic contribution (Messer 1997; Pilla and Dantas 2016; Harris-Fry et al. 2017). Anthropological and sociological research has long acknowledged relationships between food, gender, power, and interpersonal conflict within household units (Appadurai 1981; Nichter 1981; Ellis 1983; Whyte and Kariuki 1991; Chilton and Booth 2007; DeVault 2008; Counihan and Kaplan 2013). Yet with few exceptions (Cardoso et al. 2016; Fong et al. 2016), many recent analyses of household food insecurity have neglected or downplayed the role that IPV might play in driving experiences of food insecurity. This remains true despite recent global emphasis on gender-based violence research (García-Moreno et al. 2005; McIlwaine 2013), recognition of the importance of gender-based abuse to food insecurity experiences at the highest levels of the Food and Agriculture Organization of the United Nations (2017), and independent calls for "woman-centered" policies and interventions to address both IPV (Swart 2012; De Schutter 2013) and food insecurity (Van Esterik 1999).

In a previous work, I presented findings concerning the relationship between experiences of food insecurity, uncertainty, intimate partner conflict, and the subsequent instability of marital unions in Kenyan informal settlements (Gilbertson 2015). There I argued that experiences related to *njaa* (hunger) are painful and emotionally jarring, and often contribute to experiences of conflict and abuse between intimate partners—in some cases, even the dissolution of marriages and household units. In this chapter, I extend that argument to show how and why violence must also be recognized as a potential driver of household food insecurity, especially in low-resource settings like informal settlements. To do so, I will first provide an overview of the field site and methods used in this research. Next, I describe the most common types of violence or abuse in a Kenyan informal settlement. Following this, I consider the sources of intra-household conflict, and offer *pesa* (money) as the foremost reason couples fight. Subsequently, I consider the most common responses household members make in their attempts to cope with *njaa* and less severe experiences of food insecurity. Afterward, I address how experiences of conflict, abuse, and food insecurity coincide within households, including how people actively respond to conflict. Finally, I review this chapter's findings, including the causal, cyclical relationship between experiences of intra-household violence and conflict and food insecurity, and argue for women-centered interventions to address both violence and food insecurity.

Field site and methods

This chapter is based on fieldwork that took place in the informal settlement known as "Bangladesh" or "Bangla," in Mombasa, Kenya. Mombasa is Kenya's second most populated city and East Africa's most important seaport. Bangla is located in an industrial suburb on the mainland just west of Mombasa Island. At the time of fieldwork, Bangla's population was around 10,000 to 15,000 people. Many residents are economic migrants from western Kenya, who claim ethnic affinity with the Luo people. Population density varies across the settlement. Some areas are highly populated, with mud-thatched, corrugated-iron-roofed houses packed tightly together. In other areas, open green spaces produce sugarcane, maize, and other foodstuffs for local consumption. Plastic and paper refuse, burned garbage, raw sewage, and scavenging chickens, pigs, and cattle are common to this landscape. Marked inequality exists concerning resource access (e.g., water and electricity).

Fieldwork took place over 15 months between July 2009 and December 2010. The methods utilized included participant observation, 109 semi-structured interviews with 67 participants, 3 focus group discussions, and 2 questionnaires involving 28 households (Gilbertson 2013, 2015). All research activities received ethical approval by the University of Oxford. To protect participants' identities, all the names used in this chapter are pseudonyms.

Violence within Bangla's households

Spousal abuse is very common in Bangla and featured prominently in interviews and informal conversations with both women and men. While it cannot be considered representative, more than half of the 21 women I asked agreed that in the past 4 weeks, they had heard complaints from at least one friend, neighbor, or other acquaintance about intimate partner abuse. The types of abuse described by residents of Bangla closely match the typology previously enumerated from work in South Africa (Van Rensburg 2007), including sexual, physical, emotional, verbal, and financial (or economic) abuse. For instance, sexual abuse involves any forced or coerced sexual activity (García-Moreno et al. 2005). In Bangla, sexual abuse within conjugal relationships tends not to be recognized as rape because such behavior is often justified as part of a man's "marital rights." The most common examples of physical abuse include slapping, punching, biting, kicking, throwing objects, "cane crooks," poisoning, and stabbing.

Verbal abuse refers to the use of any insulting, demeaning, or otherwise hurtful language. In Bangla, this is referred to by the verb *kugombana* (to quarrel) or the noun *ugomvi* (quarrel; e.g., *ugomvi ndani nyumba*, or quarrel in the house). One participant described such behavior as speaking "harshly" or "throwing bad words on you." For instance, one woman's husband called her *malaya* (prostitute) because he doubted her commitment to him.

Emotional abuse has been defined as acts of humiliation, intimidation, and controlling behaviors such as jealousy or limiting a woman's mobility (García-Moreno et al. 2005). In Bangla, these abuses are extended to any behaviors causing frustration, anguish, and/or emotional pain. Examples include bullying, disrespectfulness, infidelity, mistrust, refusing to help out around the house, and a lack of *kujali* (caring). Others include husbands demanding to know their wives' whereabouts, telling partners with whom they are allowed to spend time, or preventing spouses from visiting their natal families.

Among married women, financial violence was one of the most commonly cited types of abuse. Financial or economic violence generally refers to a male partner preventing his wife from earning money, exploiting her financially, taking her earnings, or refusing to give her the money she needs for household expenses (Jewkes et al. 2003; Dunkle et al. 2004; Van Rensburg 2007; Postmus et al. 2011; Sanders 2015). Although this definition applies in Bangla, certain adjustments are necessary as women also use financial abuse against their husbands. Additionally, financial violence in Bangla includes denying a partner input into how to spend money, abandoning a household to its economic detriment, using money to coerce a partner, and using money to cause other harms.

While dividing abuse experiences into categories may be academically useful, doing so rarely reflects the realities of lived experiences of violence; for instance, physical violence is always also emotional violence (Phorano et al. 2005). Similarly, financial abuse may also represent physical and emotional violence, if such behavior causes anxiety or forces a spouse to go hungry. In light of this, the rest of this chapter moves away from use of abuse typologies to rely more on narrative descriptions involving abuse and its consequences.

Sources of intra-household conflict and abuse

In order to understand the relationship between violence and food insecurity, it is necessary to consider the sources of conflict within relationships. When I asked interviewees to tell me the most common reasons why they quarrel with their partners, responses included money (*pesa*) most importantly, and especially husbands' failures to provide for their families. Additionally, participants cited food insecurity, infidelity, refusing to use condoms, HIV, and gossip. Other sources included disrespecting or ignoring a partner, alcohol abuse, wives failing to meet domestic expectations (e.g., cooking, cleaning, sex without contraceptives, and childcare), wives refusing to do what husbands ask, and deceiving one's partner.

Pesa (money)

In Bangla, where "*pesa ni kila kitu*" (money is everything), including food security, it is unsurprising that money emerged as the most contentious issue within households. As one married man expressed: "Concerning money, that is a field of war"; if there is a lot of money, then you can live the good life, but "if [there is a] lack of cash in the house, it is [a] quarrel." Indeed, research from Kenya (Silberschmidt 1999, 2001) and South Africa (Gibbs et al. 2014) suggest that a man's inability to meet the expectation to provide for his family challenges his masculinity, and may provoke attempts to reassert his dominance through alcohol abuse, infidelity, and physical or sexual violence. Interviews with wives in Bangla, however, revealed that husbands' failures to provide also motivate their physical violence against men.

The use and control of money within households is a powerful, emotionally charged subject, especially among those confronted by the prospect of *njaa*. Therefore, beyond a man's failures to provide, the ways that money might be abused, and how these abuses relate to domestic power, are also vital to understanding how finances inspire conflict. For instance, a partner withholding money needed for food or other necessities is an act guaranteed to create conflict and antagonism, if not outright violence. Sharon, a kiosk-owner and tailor, described frequent quarrels with her husband, a security guard. She explained how her husband consistently contributes only a small portion of his salary to household needs (like food), choosing instead to spend most of his earnings on *chang'aa* (illicit alcohol). Sharon pointed out that despite the inadequacy of his contribution, he expects her to provide him with the nice meals he prefers. While a lack of money is an understandable, relatable reason for failing to give your partner money, refusing to give money when you have it is not. Such refusals express disrespect, contempt, and the intention to subjugate a partner. Consequently, how people budget for daily/monthly expenses reflects the status of their relationships and helps to determine contexts for future interactions.

Responses to food insecurity

Although households that are severely food insecure often have no other option but to go hungry (referred to as *kulala njaa*, or to go to sleep hungry), most people in Bangla are usually able to access some food to eat each day. To do so, members of households attempt through practical responses to cope with insecurity, and to avoid the suffering and shame of *njaa*. These include decisions, strategies, and actions aimed at managing risk with available resources. The most common responses consist of stretching budgets and/or dietary changes—adjusting what, when, where, and how much food is eaten (and by whom). This includes decreasing dietary diversity, eating less-preferred (usually less-expensive) foods, eating away from home at local informal cafés (for those with money), reducing intake (eating smaller, fewer, meals or outright skipping them), and prioritizing the needs of some members above others. Beyond these, seeking new and possibly less-preferable employment (e.g., more strenuous, hazardous, or lower-paying work), saving money secretly, and responses that call upon friends and neighbors or other social networks for assistance are alternative strategies. Additional, more desperate, less-preferred/socially acceptable responses involve working for food, selling valuables, begging, theft, and exchanging sex for money or food.

Conflict, abuse, and food insecurity

In Bangla, food insecurity and conflict are often experienced together, and a shortage of food may create or exacerbate conflict between household members, especially husbands and wives (Gilbertson 2015). The opposite dynamic often occurs as well, when intra-household conflict contributes to experiences of food insecurity. Refusing to cook, to eat, or to provide money needed for food are three ways individuals in conflict with their partners (or other household members) may use food as an instrument of abuse, as well as a tool to renegotiate power within these relationships.

Refusing to cook

In response to conflict, especially disagreements involving money and/or a husband's failure to provide, wives may refuse to cook. For women who are unequal, if not subordinated, refusing to cook offers one of the few culturally tolerated responses (alongside refusing sex) through which to express frustration and unhappiness. When food is available, women may choose to cook for themselves and their children while their husbands are away, and claim not to have eaten when the husbands return. Although most of the women I interviewed seemed convinced that such subterfuge was unknown to husbands, the men I talked with claimed awareness. According to Mzee Daniel, a married man in his seventies,

> After work, if a man goes for *pombe* [beer], the woman knows this, so she and the children eat and clean all the dishes and utensils quickly. Then, when the husband arrives home, they tell him that they are hungry and have had nothing. First, she wants her husband to know that there is nothing in the house because he hasn't contributed enough. Second, there is a lack of trust, so the woman believes that if you have money and you are not sharing it with the family then you must be spending it on other women.

Refusing to eat

When wives and husbands are in conflict, it is not unusual for one partner (often the wife) to refuse to eat. For instance, Lydia, a middle-aged woman living with HIV, told me that even if she and her husband are seated at the table and ready to eat, if they begin to quarrel either or both partners might leave without eating. The reasons why a spouse may refuse to partake in a meal vary and include conscious, deliberate acts of defiance, as well as less-intentional responses to the stress that accompanies conflict. For instance, while describing her experiences of abuse with her husband, Barnabas, Diana, a mother of four in her early twenties, made clear that following fights she would refuse to cook or eat. Explaining this, she told me,

> Maybe [...] we have quarreled. So [...] I cannot provide food on time for, maybe for the children or for the family. [...] Because I'm [...] angry. [...] Maybe we have quarreled about something, so maybe he has fired me with bad words. So I still have the pain in my heart. [...] Like the time we have quarreled, since 15th [a week before] [...] I don't take food; I take maybe *chai* [tea], only *chai* until night. So at night I take only *maziwa* [milk] [...] so I didn't take food.

In a statement that revealed the aftermath of learning about her husband's infidelity, Diana told me, "Since that stress has catched me, I don't eat much, I don't eat lunch, I eat only breakfast, maybe supper. [...] But I don't have the appetite since." [...] "Because [...] if I'm stressed, even [my] throat it doesn't like food. [...] I can't eat."

Within informal settlements, where *njaa* is a daily reality for so many, refusing to eat when food is available is a powerful expression. By choosing not to eat, and thereby making unnecessary sacrifices (and suffering because of them), a wife or husband may make clear that the marital relationship is in crisis. In addition to expressing a "lack of peace" within relationships, when husbands skip meals, or refuse to eat what their wives cook, wives attach other meanings to these acts: namely, that their husbands are unfaithful. This often leads to immediate conflict.

Refusing to give money

Following conflict, some individuals (usually husbands) may refuse to give their partners money. Mama Rukiya, the successful owner of a small informal café, told me that, "Here, food is money. So, after quarreling with husband, who has money in his pocket, and he disappears with that money. Definitely that family will go without eating." Suggesting that the intent behind such responses is to punish the partner, Robert (a day laborer whose wife lived in western Kenya), said that on mornings following a quarrel, he would not give his wife money for lunch so that she would "go without" and "suffer" for the day. He added that his motivation was to "learn her a lesson [...] not to misbehave in my house." In addition to refusing to give their wives money, husbands may use what money they have to eat outside of the household, even when their wives are willing to cook. In households where wives are breadwinners, they too may decide to refuse to contribute food in order to hurt their husbands. Recounting such an instance, Mama Rukiya told me how she sometimes refuses to buy food when her husband fails to provide. When this happens, no one eats (including the children).

In response to conflict, wives may also punish husbands by simply taking their money. According to Robert, after beating a wife, "You will go hungry because she has run away with cash. She will time you. [...] She'll just go away with that cash and make you suffer. We'll go hungry for even one week [chuckles] before seeing her." Providing further detail, Robert explained that when husbands quarrel with their wives, they beat them in the evening and then go to sleep. "And with women, they don't tell us their plans. Next day, you will go out, or when [you] go to job, and coming back, [you] find nobody is there." Noting the potential consequences for children, he added, "The children also will suffer. Because if the mother runs away with money, they will just remain there begging from neighbors."

Running away

"Running away" from a household, with or without children, temporarily or permanently, represents an alternative and more drastic response to intra-household conflict. Although running away is more commonly associated with women, men are also known to abandon households, especially if they are overwhelmed by an inability to provide. For wives with children, running away represents a response of last resort, prompted by overburdening, experiences of violence, and/or an overwhelming sense of domestic insecurity. For example, when I asked Mama Rukiya why she ran away from her husband, she told me that they quarreled all the time and that she had been frustrated and stressed. Mama Aisha, a very poor vegetable seller who looked much older than her 40 years, ran away from her husband and co-wife "*mara nyingi*" (many times) due to physical abuse. In such cases, running away may represent the only viable response. According to Diana, while a wife could report her abusive husband to the police or to village elders, doing so is very difficult: "I'm weak, I can't do that. If I do that, where will I go? I can't return [to] this house again because I might be killed."

Beyond providing respite from conflict and other burdens in the household, running away to stay with friends or one's natal family communicates to a husband how difficult domestic life is without a wife. Citing wives running away, but leaving children, as another way wives may punish their husbands, Moses, a jobless, divorced, recent arrival from western Kenya, explained,

> When a man is left with little children, and maybe he's the person who must go out in the *jua kali* [hot sun] to find something to bring back home, then it is a headache because you go out, you leave behind the little children, and maybe there is nobody to take good care of them. And as life is in the slum [...] the sewages, the environment is not conducive, so it is just a problem.

Jacob, a married man employed as a full-time watchman, expressed that losing his wife would mean adding the cost of childcare to his budget. When asked if harmony in the family is important to food security, Jacob strongly asserted it was. Noting how quarreling encourages wives to run away, Jacob explained that the loss of a wife also means the loss of any income she brought home—income likely used for food.

Secret savings

When husbands consistently fail to provide food, wives will commonly respond by keeping money that their husbands do not know about—what I have translated as *akiba ya siri* (secret savings). According to Moses, most wives keep at least a small amount of cash, regardless of their financial means or status of their relationships. In explanation, he told me: "It is common in the slums [because] money is very scarce, so women would like to save little by little. [...] So it is because of the *maskini* [poverty] and the life in the slums."

It nevertheless became clear that wives also save money secretly as a response to uncertainty in their domestic relationships, and particularly as responses informed by their inability to depend on their husbands to provide for their needs. In this way, secret savings offer a sense of security and ability to provide for oneself and one's children when the husband fails to do so, particularly if an emergency arises, or if a partner dies or leaves. Explaining why she keeps secret savings, Diana told me,

> My future, it is mixed up because every year, every time, every month, we are quarreling about flimsy things. [...] So I put that money [...] I'm feeling like a bird, which I don't know where to put my thing which can help in life. Maybe that money can help me; anything can happen in our family.

Saving secretly is thus inspired as much by a husband's past failures (and refusals) to provide and a wife's experiences (and expectations for future experiences) of food insecurity, as it is by experiences and future expectations of violence. Both Diana and Mama Rukiya made this clear by admitting to saving money in order to provide themselves with the option to run away if needed. Emphasizing saving secretly as a response to conflict and uncertainty, Moses told me,

> In the slum life, the way we live, most women want to have the savings because anything can happen. There can be a quarrel; and, maybe she knows that if [she's] got a saving, and if there arises a quarrel, [she] can have a way to go home [to her natal family].

Conversely, husbands tend to perceive secret savings as a source of risk in the household. This is because a wife with money is not (or is less) dependent on her husband, and may be harder for him to control. As Benjamin, a shop owner, landlord, and 25-year resident of Bangla, put it, "When a woman is having money, is a problem. [...] Because she'll have to decide on what to do at any time without consulting her husband." Additionally, secret savings suggest a lack of unity in the household, as if a wife is "hedging her bets" on the future viability of the marriage. Highlighting the reasons men are threatened by their wives' savings, including that women can use them for "family planning" (including abortions), Moses explained: "You know, the secret money being kept, it do have some ill motives. The woman may save so that she may escape when you are not aware. You just come back in the evening, and you find the room empty." Barnabas, a 47-year-old landlord who rented mud-thatch rooms to some of Bangla's poorest residents, explained prevailing attitudes of men whose wives have money: "Some [men] think their wives will grow horns" and "will become rude or develop powers that they didn't have before." As a result, some men try to keep their wives from working to ensure that they have no savings. Barnabas emphasized that a working wife can reflect badly on a husband in the community, and that it can make him feel "inferior," especially if the work involves "odd jobs" such as "weeding" or "carrying water for other women." Barnabas noted how jealousy also plays into these dynamics, with some husbands thinking that, by his wife working, "she is getting openings, chances to become promiscuous." This helps to explain why, if they discover it, some husbands will take this money from their wives, or will try to force them to spend it all on food or other household necessities—actions guaranteed to contribute to further conflict.

Infidelity

In response to food insecurity at home, some women decide to seek new partners in order to establish alternative means of support. However, spousal conflict also elicits this response since abused, disrespected, or otherwise mistreated women may be more inclined to stray from their marriages, sometimes out of revenge or a desire to find safety and/or solace in a new relationship. Unsurprisingly, suspected or confirmed infidelity is a potent source of conflict within marriages; an unfaithful wife runs the risk of divorce, physical violence, and other abuses—including abuses that may exacerbate food insecurity.

According to Mama Rukiya, when men turn to infidelity, they do so often as a response to conflict with their wives. Subsequently, and intentionally or unintentionally, husbands may put the food security of their households at risk by spending desperately needed money on these other sexual partners. Echoing these sentiments in an explanation of why men may become unfaithful as a response to intra-household conflict, as well as how that infidelity may put their families (especially children) at greater risk of hunger, Moses offered,

> Once there is conflict in the house, that one will frustrate the mind of a man. And even if he gets some money, he must go and drink because he

> just see[s] that as a solution. Maybe after drinking, he will come home drunk and the wife will not bother him so much. So in the process of going out to drink, he will reduce that money[...] and the problem will continue to breed. [...] Most men do go out [have "affairs"] because they just want somewhere where they can go and have that peace of mind. Yeah, so conflict in a house can cause food shortage in that way.

Cycles of conflict, abuse, and food insecurity, and the impact of women-centered intervention

Providing for the needs of a household on a daily basis in an informal settlement often requires coordinated efforts involving the hard work of two or more individuals, especially the breadwinner and the person who prepares meals. Any factors that hinder cooperation, especially between husbands and wives, have the potential to increase food insecurity. An analysis of findings from Bangla reveals three important points that link conflict between husbands and wives to food insecurity, and a fourth that suggests that women-centered interventions are necessary in order to reduce global food insecurity.

First, experiences of conflict have the potential to distract household members from normal, everyday, domestic, and wage-earning activities that are necessary to ensure that members of a household are able to eat. According to Vincent, a local health coordinator employed in Bangla, "Once quarreling starts, everything comes to a standstill in a household." Likewise, Lydia, a *chang'aa* brewer and 32-year-old woman living with HIV observed, "Some people in some houses, they fight and forget about the children. They don't provide any food, so the kids are outside crying." Similarly, Diana offered,

> Like now, the food is there, that *githeri* [maize and bean stew], but I could not be able to arrange to prepare it until now because of the quarrels since morning, I used to put that *githeri* at night so if I wake up in the morning, it is ready, I try it. But quarrels, since yesterday until morning, so it has [made] me lose even interest of doing such-a-things.

Children, elderly grandparents, and other household dependents go hungry when a mother cannot cook because she must flee to a neighbor's house to avoid abuse, or when a father is too distraught to work, or if another breadwinner (e.g., a teenager or adult child) must care for their siblings instead of earning a wage.

In addition to distracting household members, experiences of conflict have disintegrating effects on household relationships, leading people to put their own best interests above those of others. For instance, when I asked Mama Rukiya about issues of food and spousal conflict, she explained, "The house will have no peace. [...] Everyone will be thinking of his or her own things separately." As relationships fray and become adversarial, and spouses (or other household members) seek to punish each other for perceived transgressions, it becomes more difficult to mobilize unified efforts to ensure that food is available and cooked on time for meals. Individual decisions and actions taken in response to conflict represent the means through which conflict may create or (more likely) exacerbate existing conditions of food insecurity. Violence and abuse within a household make it less food secure, and what matters most is how household members *respond* to this conflict.

Second, the responses to conflict that present the most risk are those made unilaterally, in secret, and without the consent of others. For instance, a husband's decision to eat away from home (when there is little or no food at home), or a wife's decision to cook, eat, and wash-up before her husband arrives home, are both responses that are likely to generate further conflict. In these cases, any improvement to an individual's food security comes at the expense of others' well-being. While some responses may affect food security directly (e.g., refusing to cook or to give money, or running away), others such as infidelity may do so indirectly by inspiring further (or intensifying current) experiences of conflict in the household. Paradoxically, responses that are intended to improve a household's future food security, such as a wife saving money secretly, may have the opposite effect: if a husband discovers these savings, conflict will result and long-term food security may decrease. Similarly, a husband's decision to eat outside of the household does not necessarily put any other household member at risk (if food is available at home), but the infidelity this suggests may have an unintended consequence. In these cases, responses to conflict or insecurity only introduce risks after they are discovered, interpreted, and assigned meaning by others in the household. In this way, the risks associated with responses are contingent on interpretation and the current condition of the pertinent relationships. Each of these responses may be interpreted as a strong indication of uncertainty and/or a lack of commitment to the household and to the well-being of its members. As a result, these responses may inspire conflict and/or introduce further risk through secondary responses, such as a wife's refusal to cook or a husband's withholding of money.

Third, since experiences of food insecurity predispose household members to conflict, and conflict may incline members to further food insecurity, it seems reasonable to conceptualize the relationship between household food insecurity and conflict as dialectically linked and cyclically perpetuating. As Moses put it, "But I think they work vice versa. You know, if conflict causes a lack of food, then it is that lack of food that will breed more conflict. So it works that way." Given the right conditions, either food insecurity or interpersonal conflict (abuse, violence, etc.) may encourage the emergence of the other—although in some cases, and from certain perspectives, the emergence of food insecurity and conflict are experienced simultaneously.

Fourth, findings from Bangla suggest that global efforts against hunger and food insecurity cannot succeed unless they also address the concurrent epidemic (García-Moreno et al., 2005) of intimate partner violence (IPV). To do so, governments, policy makers, non-governmental organizations, nonprofits, social enterprises, community-based organizations, international funders, and other stakeholders must adopt women-centered approaches that place foremost the notion that household food insecurity and IPV are inextricably related issues. These approaches and interventions should have at their core the enhancement of female health, well-being, and empowerment. Encouragingly, recent studies have shown that cash and in-kind transfers of food to insecure households in Ecuador reportedly reduced daily stress levels and experiences of IPV among spouses (Buller et al. 2016; Hidrobo et al. 2016). My research suggests that the opposite may also prove true: interventions that are effective at reducing violence within households, especially abuse between intimate partners, may increase food security.

The observation that there are few studies that have made IPV a focus in food security research is perplexing given the almost universal association between women and food in the household, and the likelihood that any abuse directed toward women could produce insecurity by hindering food-related activities. In this direction, further research is needed but, as a starting point, pertinent stakeholders must see food insecurity as a feminist (Phillips 2009), human rights issue (Ayala and Meier 2017) in which "women are both vulnerable and powerful, victimized and empowered through food" (Van Esterik 1999). To conceptualize food security otherwise may be to ignore the importance of the sociocultural, political, and economic relationships and interactions found within food insecure households.

The findings presented in this chapter support an interpretation of intra-household dynamics in which conflict may be seen as an almost unavoidable outcome of experiences of chronic and severe food insecurity, especially involving *njaa* (Gilbertson 2015), but increased food insecurity and hunger are never inevitable outcomes of conflict between intimate partners. Reductions in food security are only potential outcomes of conflict, since the realization of these outcomes is dependent on many factors, including individual circumstances and household members' access to income and savings. Certainly, if a household is relatively prosperous, and food is not a worry, it is less likely that conflict will cause *njaa*. That said, even within well-off households, breadwinners could still refuse to bring home food or to provide the money needed to purchase it, wives may refuse to cook, and some household members may refuse to eat. When any of these responses come into play, it is likely that some household members will go hungry (especially children and other vulnerable dependents).

Conclusion

Experiences related to food, cooking, and eating in the household have the potential to contribute to a shared sense of security, goodwill, and harmony, or to foster friction, mistrust, insecurity, and uncertainty. Household food insecurity and conflict are intimately related phenomena. As a result, the indispensable relationships people hold most dear—relationships that are essential to managing food insecurity and avoiding *njaa* through "coping" responses—also introduce risk into their lives. Therefore, while it seems clear that household units may function as risk-management mechanisms that "share" or pool risks associated with income and production among multiple members (Doss 2001), it seems equally clear that when it comes to conflict, especially between husbands and wives, intra-household relationships must also be considered as potential sources of risk.

People respond to food security risks through the choices, actions, and resources available to them. But these responses are often severely constrained by poverty. In such contexts, resources including money, food, and access to employment/income (as well as risks such as violence), tend to be unevenly distributed. As a result, the challenges to reducing food insecurity are complicated, and not simply a matter of alleviating poverty and facilitating household access to food. Instead, improving food security in poor urban communities like informal settlements is equally an issue of negotiating stability and equality within intra-household, especially intimate, relationships.

References

Appadurai, Arjun. 1981. "Gastro-politics in Hindu South Asia." *American Ethnologist* 8(3): 494–511.

Ayala, Ana and Benjamin M. Meier. 2017. "A human rights approach to the health implications of food and nutrition insecurity." *Public Health Reviews* 38(1): 10.

Buller, Ana M., Melissa Hidrobo, Amber Peterman, and Lori Heise. 2016. "The way to a man's heart is through his stomach?: A mixed methods study on causal mechanisms through which cash and in-kind food transfers decreased intimate partner violence." *BMC Public Health* 16: 488.

Cardoso, Lauren F., Jhumka Gupta, Sara Shuman, Heather Cole, Denise Kpebo, and Kathryn L. Falb. 2016. "What factors contribute to intimate partner violence against women in urban, conflict-affected settings? Qualitative findings from Abidjan, Cote d'Ivoire." *Journal of Urban Health* 93(2): 364–378.

Chilton, Mariana and Sue Booth. 2007. "Hunger of the body and hunger of the mind: African American women's perceptions of food insecurity, health and violence." *Journal of Nutrition, Education, and Behavior* 38(3): 116–125.

Counihan, Carole M. and Steven L. Kaplan. 2013. *Food and Gender: Identity and Power*. London: Routledge.

Crush, Jonathan and Bruce Frayne. 2010. *The Invisible Crisis: Urban Food Security in Southern Africa*. Urban Food Security Series No. 1. Queen's University and AFSUN: Kingston and Cape Town.

Crush, Jonathan and Bruce Frayne. 2011. "Urban food insecurity and the new international food security agenda." *Development Southern Africa* 28(4): 527–544.

De Schutter, Olivier. 2013. *Gender Equality and Food Security: Women's Empowerment as a Tool Against Hunger*. Mandaluyong City, Philippines: Asian Development Bank.

Devault, Marjorie. 2008. "Conflict and deference." In *Food and Culture: A Reader*. 2nd ed. C. Counihan and P. Van Esterik (eds.). 240–258. London: Routledge.

Doss, Cheryl R. 2001. "Is risk fully pooled within the household? Evidence from Ghana." *Economic Development and Cultural Change* 50(1): 101–130.

Dunkle, Kristin L., Rachel K. Jewkes, Heather C. Brown, Glenda E. Gray, James A. McIntyre, and Siobán D. Harlow. 2004. "Gender-based violence, relationship power, and risk of HIV infection in women attending antenatal clinics in South Africa." *The Lancet,* 363(9419): 1415–1421.

Ellis, Rhian. 1983. "The way to a man's heart: food in the violent home." In *The Sociology of Food and Eating.* A. Murcott, (ed.). Aldershot, UK: Gower Publishing.

Faye, Ousmane, Angela Baschieri, Jane Falkingham, and Kanyiva Muindi. 2011. "Hunger and food insecurity in Nairobi's slums: An assessment using IRT models." *Journal of Urban Health* 88(Suppl. 2): 235–255.

Fong, Sarah, Jhumka Gupta, Denise Kpebo, and Kathryn Falb. 2016. "Food insecurity associated with intimate partner violence among women in Abidjan, Cote d'Ivoire." *International Journal of Gynecology & Obstetrics* 134(3): 341–342.

Food and Agriculture Organization of the United Nations. 2017. *Gender-based Violence Affects Food Security and Nutrition: Interview with FAO's Deputy Director-General, Daniel Gustafson.* Accessed July 11, 2018, www.fao.org/fao-stories/article/en/c/1069792/.

García-Moreno, Claudia, Henrica. A. F. M. Jansen, Mary Ellsberg, Lori Heise, and Charlotte Watts. 2005. "WHO multi-country study on women's health and domestic violence against women: initial results on prevalence, health outcomes and women's responses." Geneva: World Health Organization.

Gibbs, Andrew, Yandisa Sikweyiya, and Rachel Jewkes. 2014. "'Men value their dignity': Securing respect and identity construction in urban informal settlements in South Africa." *Global Health Action* 7(1).

Gilbertson, Adam. 2013. *The Ecology of Risk in an Informal Settlement: Interpersonal Conflict, Social Networks, and Household Food Security.* Dissertation for D.Phil., University of Oxford.

Gilbertson, Adam. 2015. "Food security, conjugal conflict, and uncertainty in 'Bangladesh', Mombasa, Kenya." In *Ethnographies of Uncertainty in Africa.* E. Cooper and D. Pratten (eds.). New York: Springer.

Harris-Fry, Helen, Niva Shrestha, Anthony Costello, and Naomi M. Saville. 2017. "Determinants of intra-household food allocation between adults in South Asia—A systematic review." *International Journal for Equity in Health* 16(1): 107.

Hidrobo, Melissa, Amber Peterman, and Lori Heise. 2016. "The effect of cash, vouchers, and food transfers on intimate partner violence: Evidence from a randomized experiment in Northern Ecuador." *American Economic Journal: Applied Economics* 8(3): 284–303.

Jewkes, Rachel. K., Jonathan B. Levin, and Loveday A. Penn-Kekana. 2003. "Gender inequalities, intimate partner violence and HIV preventive practices: Findings of a South African cross-sectional study." *Social Science & Medicine* 56(1), 125–134.

K'akumu, Owiti A. and Washington H. A. Olima. 2007. The dynamics and implications of residential segregation in Nairobi. *Habitat International* 31(1), 87–99.

Kimani-Murage, Elizabeth W., Lilly Schofield, Frederick Wekesah, Shah Mohamed, Blessing Mberu, Remare Ettarh, Thaddaeus Egondi, Catherine Kyobutungi, and Alex C. Ezeh. 2014. "Vulnerability to food insecurity in urban slums: Experiences from Nairobi, Kenya." *Journal of Urban Health: Bulletin of the New York Academy of Medicine* 91(6): 1098–1113.

McIlwaine, Cathy. 2013. "Urbanization and gender-based violence: exploring the paradoxes in the Global South." *Environment and Urbanization* 25(1): 65–79.

Messer, Ellen. 1997. "Intra-household allocation of food and health care: Current findings and understandings—Introduction." *Social Science & Medicine* 44(11): 1675–1684.

Nichter, Mark. 1981. "Idioms of distress: Alternatives in the expression of psychosocial distress: A case study from South India." *Culture, Medicine and Psychiatry* 5(4): 379–408.

Phillips, Ruth. 2009. "Food security and women's health: A feminist perspective for international social work." *International Social Work* 52(4): 485–498.

Phorano, Odireleng, Keitseope Nthomang, and Dolly Ntseane. 2005. "Alcohol abuse, gender-based violence and HIV/AIDS in Botswana: Establishing the link based on empirical evidence." *SAHARA-J: Journal of Social Aspects of HIV/AIDS* 2(1): 188–202.

Pilla, Loretta and Jaya A. R. Dantas. 2016. "Intra-household nutritional dynamics: A cross-sectional study of Maasai communities in Kenya." *Qualitative Health Research* 26(6): 793–806.

Postmus, Judy L., Sara-Beth Plummer, Sarah McMahon, N. Shaanta Murshid, and Mi Sung Kim. 2011. "Understanding economic abuse in the lives of survivors." *Journal of Interpersonal Violence* 27(3): 411–430.

Sanders, Cynthia K. 2015. "Economic abuse in the lives of women abused by an intimate partner: A qualitative study." *Violence Against Women* 21(1): 3–29.

Silberschmidt, Margrethe. 1999. "'Women forget that men are the masters': Gender antagonism and socio-economic change in Kisii District, Kenya." Uppsala: Nordic Africa Institute.

Silberschmidt, Margrethe. 2001. "Disempowerment of men in rural and urban East Africa: implications for male identity and sexual behavior." *World Development* 29(4): 657–671.

Swart, Elizabeth. 2012. "Gender-based violence in a Kenyan Slum: Creating local, woman-centered interventions." *Journal of Social Service Research* 38(4): 427–438.

United Nations Human Settlements Programme. 2016. "Slum Almanac 2015/2016: Tracking improvement in the lives of slum dwellers." Nairobi: UN Habitat.

Van Esterik, Penny. 1999. "Right to food; right to feed; right to be fed. The intersection of women's rights and the right to food." *Agriculture and Human Values* 16(2): 225–232.

Van Rensburg, M. S. Janse. 2007. A comprehensive programme addressing HIV/AIDS and gender based violence. *SAHARA-J: Journal of Social Aspects of HIV/AIDS,* 4, 695–706.

Whyte, Susan R. and Pauline W. Kariuki. 1991. "Malnutrition and gender relations in Western Kenya." *Health Transition Review* 1(2): 171–187.

3 Food is a gift of the earth

Food sovereignty among migrant farmworkers in rural Vermont

Jessie Mazar and Teresa Mares

Introduction: conducting food-security research in Vermont

While walking through the garden with Tomás in summer 2014, we were surprised to learn that he was growing a variety of *epazote* that we had not yet seen in Vermont. When asked where he found the seeds for this traditional culinary herb, Tomás explained that someone from his village in Mexico sent him the seeds in the mail. On this warm summer day, we stood with Tomás behind the trailer, where he lived with two other Mexican farmworkers, next to the farm where they worked. Across the dirt road, the corn was at full height, and there was a lingering, almost sweet, smell of manure in the air. We were in the northwestern part of the Green Mountain State—dairy country. As Tomás gave us the backstory of these seeds, we also learned about the rural village where he was raised in the state of Guerrero. There were no cars in his village and, in order to reach the nearest city, Tomás and the other villagers had to walk for twelve hours to get to a main road, where they could catch a bus. And so, as is common in many rural Mexican villages, they grew all their own food and raised their own animals. As we listened to this story, we were reminded how far Tomás was from home.

The garden that Tomás grew in Vermont was a small space of autonomy, yet it was surrounded by a landscape of anxiety and fear. As year-round employees, migrant farmworkers in the dairy industry do not qualify for temporary H2A work visas, resulting in an estimated 90 percent of the approximately 1,200 farmworkers living and working in Vermont without legal documentation. These citizenship barriers leave migrant farmworkers like Tomás in a compromised and vulnerable position when they are in public spaces and, at times, even when they are home or at work. As Vermont is one of the whitest states in the nation, farmworkers from Mexico and other Latin American nations are particularly visible in the rural areas where dairy farms dominate the working landscape. This hypervisibility in public spaces leaves farmworkers particularly vulnerable in Vermont because of its international border with Canada and the resulting heightened surveillance of border patrol and Immigration and Customs Enforcement (ICE) personnel (McCandless

2010; Mares 2019). These problematic circumstances have salient impacts on food access options and strategies among the migrant farmworkers, who are currently sustaining Vermont's agricultural economy.

More than eight years ago, we set out to understand a problematic contradiction in the US food system and how it might manifest in Vermont—the fact that the people who are feeding the nation are experiencing high levels of food insecurity themselves. We have found that studying food security among farmworkers in Vermont is deeply complicated given the highly variable and ever-changing political, social, and economic conditions in which they live. In this chapter we explore the complexity of defining and measuring food insecurity among migrant farmworkers in Vermont. We argue that the current food security measurement tool, the USDA Household Food Security Survey Module (HFSSM), falls short in capturing the lived experiences that we have observed and recorded through our research. Defining and measuring food security, particularly in cross-cultural settings, is a contentious subject among researchers. Undocumented farmworkers in Vermont encounter food access obstacles that exceed financial barriers, including minimal access to culturally familiar foods, lack of transportation, fear of going to the grocery store, and busy work schedules (Mares 2019). Further, the current classification of a household in the food security measurement tool does not allow for an understanding of the experiences of migrant workers, who are often supporting households in other parts of the United States and Mexico.

At the same time, we have also observed migrant farmworkers and their families navigating these barriers with resilience and creativity in the ways that households access food and the agency they exert over both their sustenance and cultural identities. In this chapter, we discuss another framework for understanding food access: food sovereignty. Building upon seven years of applied ethnographic research, we examine food access strategies employed by migrant farmworkers living and working in the northern borderlands; these strategies include an applied gardening project (called Huertas, kitchen garden in Spanish), mobile food *tiendas* selling culturally familiar foods, and other creative methods of food procurement.

For this chapter we draw upon our collaborative ethnographic fieldwork, which includes over eight hundred hours of participant observation in addition to in-depth interviews with twenty-five farmworkers and ten community stakeholders. Additionally, we draw upon the results of a hundred surveys we conducted with migrant households using the HFSSM. The majority of our research has been connected to and facilitated by our work as co-directors of Huertas, a community-based food access project that facilitates the planting of kitchen gardens with farmworkers like Tomás. Through sharing these stories, we strive to convey human ingenuity while also recognizing and identifying the policies and paradigms that regulate and limit human autonomy.

Encerrado: migrant farmworker (in)visibility in Vermont's landscape

In order to understand the complexity of measuring and defining food security among Vermont's migrant farmworkers, we must first discuss the geopolitical landscape. Dairy production is an essential part of Vermont's agricultural economy, accounting for approximately 70 percent of the state's agricultural sales, making Vermont the most dairy-dependent state in the nation. This production supplies socially and environmentally conscious companies, such as Ben & Jerry's and Cabot, and reinforces the state's branding of the working landscape that families have tended for generations. Missing from this narrative of the quaint Vermont countryside is attention to the presence of, and to the reliance on, undocumented laborers from Mexico and other parts of Latin America. The invisibility of Vermont's migrant farmworkers within the representation of the working landscape is no accident.

The absence of a year-round visa option limits the choices of dairy farm owners in hiring foreign-born workers, encouraging employers to look the other way when hiring workers who are likely to be undocumented. The proximity of dairy farms to the Canadian border puts farmworkers at risk of detention and deportation, but it also puts their employers at risk because of the heightened surveillance by US border patrols. The region hundred miles south of the US–Canada border falls under federal jurisdiction, and because the state is 157 miles long, north to south, this zone includes 90 percent of Vermont residents. Nevertheless, the 25-mile region along the border is called the "primary operating domain," where the presence of the border patrol and enforcement is higher. After 9/11, the number of border patrol agents and surveillance ops, and the general militarization of the region, dramatically increased on the northern border; Peter Andreas calls this transition the "Mexicanization" of the US–Canada border (Andreas 2005). The increased surveillance along the border causes extreme isolation for Latino/a migrant farmworkers and forces them to remain invisible, particularly the closer they live to the US–Canada border.

This heightened surveillance reinforces the necessity of farmworkers to remain out of sight on the farms where they work so as to not jeopardize their livelihoods or those of the farm owners. Susannah McCandless (2010: 217) documents the "carceral landscape" of Vermont for Latino/a farmworkers in her doctoral research; she describes how farm owners routinely go out of their way to ensure that employees can remain on the farm—often buying groceries and running errands for their employees. This feeling of being imprisoned by and within Vermont's working landscape has been repeatedly articulated by the farmworkers we have interviewed. The term that has come up repeatedly is *encerrado*, which translates to "locked in," "enclosed," or "penned up."

The level of seclusion for farmworkers is unique in the northern borderlands, as we were reminded after a 2014 visit to Vermont by Eric Holt-Giménez, an agroecologist, researcher and lecturer based in California. After accompanying us to visit farmworkers who participate in the Huertas project, he published a blog post on the *Huffington Post* site titled "Vermont's Invisible Farmworkers." He writes,

> I had never seen such grinding solitude. Living in the shadows of Vermont's local food mecca, these men and women suffer a structural violence that strikes at the core of who they are as a people. I shudder to think of their isolation during winter months.
>
> (Holt-Giménez 2014: 1)

This extreme fear and isolation have a salient impact on food access for farmworkers, and one would presume that it would be reflected in the food security data collected through the HFSSM.

Tomás, for example, who loved preparing elaborate meals for visitors and his housemates, had to rely on his boss or other community members to do his grocery shopping. This isolation marks a harsh existence for Tomás and other farmworker households living and working in the northern borderlands. While their labor provides for the foods that dominate the dairy aisles of our grocery stores, these workers are barred from entering those very stores to purchase food for themselves.

Borders have a tangible impact on the agency undocumented Latino/a farmworkers have over their mobility, access to basic needs, and overall well-being. The violence and sociopolitical history of the US–Mexico border has been documented by many scholars, including the works *Borderlands/La Frontera: The New Mestiza* (by Gloria Anzaldúa); *Fresh Fruit, Broken Bodies* (by Seth M. Holmes); and *The Land of Open Graves* (Jason DeLeón 2015). While the history, ecosystems, and cultures between the southern and northern borders are distinct, both borders produce a landscape of fear and dehumanization. The northern borderlands have remained largely understudied, and our research seeks to highlight both the harsh reality as well as the creativity and resiliency in food access strategies among Latino/a migrant farmworkers.

Reconsidering and reconciling food (in)security

Food security is defined by the USDA as "access by all people at all times to enough food for an active, healthy life" (Coleman-Jensen et al. 2017). The dominant tool for measuring food (in)security in the United States is the HFSSM, a standardized, rapid questionnaire comprising 16 questions that focus on whether a household has enough money to meet its food needs. The HFSSM was designed to capture respondents' perceptions of four domains of food security: anxiety about a household's food supply, perceptions about the inadequacy of the quantity or quality of a household's food supply, reduced adult food intake, and reduced child food intake. Through multiple stages of questions that determine the severity of food security among adults in the household, the HFSSM score results ranging from "High or Marginal Food

Security," "Low Food Security," and "Very Low Food Security" with parallel rankings for the food security rates of households with children. This methodology and its metrics has been validated by the FAO (2016) and adopted for the Food Insecurity Experience Scale (FIES) as the best choice for comparing food insecurity internationally and interculturally. Mark Nord from the Economic Research Service (ERS) worked on both the HFSSM and the FIES. Although the HFSSM provides a quick quantitative measurement of a household's food security, we have found that its emphasis on quantitative yes/no and multiple-choice questions does not capture the qualitative dimensions or the lived and culturally embedded experiences connected to food security. Our critique of the HFSSM echoes those raised by other scholars.

Drawing upon her ethnographic work with Inuit communities in the Canadian Arctic, Elspeth Ready shares important findings about the poor fit of the HFSSM model for indigenous communities who engage in mixed economies, arguing that "we need to do a better job of evaluating how etic concepts such as food insecurity relate to the lived experience of food insecure people" (Ready 2016:277). The matter of the cultural relevance of food security measurement tools is also emphasized in the work of Jennifer Coates and colleagues, who question the possibility of a common "core" of household food insecurity that transcends the boundaries of culture that could be used as the basis of a measurement tool. While suggesting that there are shared domains in food security in these cross-cultural studies (including uncertainty/worry about food, inadequate food quality, insufficient food quantity), the authors argue that the "US HFSSM (and therefore translations of the US HFSSM) does not adequately represent all potentially important domains and subdomains of the food insecurity experience" (Coates, Webb, and Houser 2003: 1446). As we have seen in our research, the HFSSM does not always articulate with farmworkers' emic understanding of food security or accurately measure how they engage in mixed economies (including gardening) and transnational economic exchange.

Anthropologists in particular have investigated the more nuanced and qualitative experiences of food security within Latino/a and Hispanic households, as documented in Megan Carney's 2015 book, *The Unending Hunger: Tracing Women and Food Insecurity Across Borders*; in the work by Rajanagandha Dhokarh and colleagues, "Food Insecurity Is Associated with Acculturation and Social Networks in Puerto Rican Households"; and finally, as seen in David Himmelgreen and colleagues, " 'I Don't Make the Soups Anymore': Pre-to Post-Migration Dietary and Lifestyle Changes among Latinos Living in West-Central Florida." Together, these scholars show in keen ethnographic detail that questions of food security within Latino/a communities and households are inseparable from considerations of cultural meaning, gendered labor, and ethnic/racial identity.

Since the development of the Spanish-language version of the HFSSM by Harrison et al. (2003), research has shown that Latino/a immigrant farmworker communities experience rates of food insecurity significantly higher than the national average, and also higher than the rates for Hispanic households more generally speaking. The most recent figures collected by USDA researchers show that 12.3 percent of people in the United States experienced food insecurity at some point in 2016, with nearly 4.9 percent experiencing "very low food security." For Hispanic households, the incidence of food insecurity in this same year was 18.5 percent, with 5.8 percent experiencing very low food security (Coleman-Jensen et al. 2017). The studies on farmworker food security cited in the next paragraph demonstrate that farmworkers experience food insecurity between three and four times the current national average with a disproportionate number reporting "very low food security" (Mares, Wolcott-MacCausland, and Mazar 2013).

Research on farmworker food security has primarily focused on seasonal workers in states with longer histories of hiring Latino/a migrant farmworkers as compared to Vermont (Essa 2001; Quandt et al. 2004; Cason et al. 2006; Weigel et al. 2007; Wirth et al. 2007; Moos 2008; Borre et al. 2010; Kilanowski and Moore 2010; Kresge and Eastman 2010; Brown and Getz 2011; Sano et al. 2011; Kilanowski and Lin 2013; Minkoff-Zern 2012; Wadsworth et al. 2016). In Vermont, prior to our work there was no published research on food (in)security among Latino/a migrant farmworker populations. Within the scholarship on farmworker food security, however, there are four studies that are methodologically and theoretically resonant with our own.

Brown and Getz (2011) base their analysis of farmworker food insecurity in California upon data gathered through the Fresno Farmworker Security Assessment, paying particular attention to the factors that produce hunger. Of the 454 respondents, 45 percent were food insecure, including 11 percent experiencing food insecurity with hunger. Their article is instructive as it views hunger and food insecurity not as natural phenomena, but rather as products of systematic inequalities. In another article based upon this same data set, Wirth et al. (2007) add further depth to these findings, demonstrating that the prevalence of hunger was lower in the summer months and that overall, factors associated with food security included income, documentation status, migration status, accompaniment status (whether someone was living with a spouse and/or children), and the utilization of food stamps.

From surveys of migrant workers in California's Salinas Valley, Kresge and Eastman (2010) report findings on food security issues, vegetable and fruit consumption, and the workers' interest in growing their own food and obtaining nutrition education. This study found that 66 percent of migrant farmworkers were food insecure, 39 percent had used food stamps, and 37 percent were already growing fruits or vegetables. Of those who were not growing their own food, 71 percent were interested in doing so. This study is significant because it goes beyond quantitative measures of food insecurity to identify the strategies that farmworker households engage in to access food, and the desire to pursue alternative strategies in accessing and preparing foods, such as gardening.

Minkoff-Zern (2012) argues against the assumption that dietary-health issues result from a lack of knowledge about procuring and preparing healthy foods. Based upon ethnographic work with indigenous Oaxacans in the northern Central Coast of California, she claims that food assistance providers tend to ignore the nutritional and agricultural knowledge held by farmworkers as well as the larger systemic issues that limit farmworkers' access to food, including structural racism and classism. This study is helpful for unpacking the structural inequalities that are embedded and perpetuated within the food system and for illuminating the diverse knowledge systems that migrants bring with them into the United States.

Finally, guided by an ecological framework, Sano and colleagues (2011) examine how low-income rural Latino/a immigrant families attempt to meet their food needs, describing both their successes and failures. The study found that a higher incidence of food security was associated with higher incomes, better financial management skills, legal documentation status, stable employment, available social support, and access to health insurance and home ownership. The ecological model that the study employs is useful in its description of the various systems that the mothers engage in feeding their families—ranging from microsystemic factors within the household to macrosystemic factors like federal laws and policies. In this way, this study is instructive in connecting an analysis of household food access with a systems-oriented approach.

Together, these studies provide a useful theoretical and methodological platform to evaluate the experiences of food security of Latino/a migrant workers in Vermont. A key point of departure between our research and these existing projects is that Vermont presents a geographic and social context very different from areas like California with long histories of Latino/a migration. Additionally, because dairy work is year-round work, there are different patterns of seasonality and temporality that shape the food access strategies of dairy farmworkers. After years of ethnographic research and participant observation, we knew that Latino/a migrant farmworkers in Vermont had unique challenges to food access, whether because of their lack of access to agricultural visas or their isolation, border proximity, and busy work schedules. As we set out to determine baseline rates of food insecurity, we expected that the results of farmworkers in Vermont would echo previous studies of farmworker food insecurity. We also expected that the survey module might not capture the lived reality of food access given our observations over years of working with farmworkers: that having enough money, which is the main indicator in the HFSSM, was not always the main barrier to accessing food. The rapid 16-question survey only asks if a household has enough money to access healthy nutritious food and does not record other barriers such as transportation or fear of being in public.

After we conducted a hundred surveys with farmworkers using the HFSSM, our results showed that 18 percent were classified as food insecure. Data from the USDA food security surveys revealed that in central Vermont's

Addison county, 15.7 percent of farmworkers are food insecure (n = 51), while among those living in northern Vermont's Franklin County (which falls in the "primary operating domain" near the federal border), 18.2 percent were found to be food insecure (n = 33). This discrepancy speaks to the challenge of accessing food, particularly in US the northern border regions.

The overall percentage, however, is much lower than previous studies reported, which found up to 60 percent of farmworkers were food insecure. As we attempt to reconcile the differences in these data, some answers emerge. First, dairy work is desirable employment for farmworkers because of the stability it can offer. In-depth interviews with farmworkers (including Tomás), who have worked other seasonal jobs around the United States, affirm that dairy work is preferred to other employment because the work is year-round and offers a higher wage than most seasonal work. Also, farmworkers are often provided housing, which is a significant challenge confronting seasonal workers. But the fact that migrant farmworkers in Vermont are quantitatively experiencing slightly elevated rates of food insecurity, while our qualitative data shows that they are experiencing significant obstacles in accessing food, indicates that there is a need to examine the metrics used to measure food access. This mismatch in quantitative and qualitative data comes back to the design and scope of the HFSSM itself.

Our quantitative data showed that a lack of a year-round visa and, therefore, living without legal documentation, and the proximity to the US–Canada border impacted food access among Latino/a migrant farmworkers in Vermont. Often, when filling out the questionnaire, respondents reported having enough money to buy the food that they need, but said they were unable to go to the store because they were afraid of *la migra* (border patrol), or they did not have transportation, or their work schedules prevented it. The survey module cannot capture this feedback, because it only asks about the financial dimensions of food security. Further, the module classifies a "household" as those who are living together under one roof. But for migrant farmworkers, the household is often not restricted to the physical spaces where they reside in Vermont. While many farmworkers live with other workers, they also support households back home. And so, while they may have enough money to purchase food for themselves, they are often sending large portions of their income to Mexico to support expenses related to healthcare, food, housing, and education for their families. Within one "household" there are often three or four young single men who are living together, but who do not necessarily share money, food, or resources with one another. The HFSSM is unable to capture the complexity of these household structures, which are often extended in the case of transnational migration.

There is a clear mismatch between the survey instrument developed to measure food security and these lived experiences. This reflects a deeper conceptual problem within food security. Raj Patel offers insight into the shortcomings of food security by discussing the paradigm that views food as a commodity versus food as a right. In an interview, Patel (in Stoneman 2010) eloquently sums up this predicament:

> No one disputes the importance of sufficient, safe, and nutritious food to lead a healthy life, the key word here is 'access." Under food security, the question of power never comes up—as long as access is guaranteed under some system or other, there's no problem. The trouble, however, is this: You can be food secure under a dictatorship. You can be food secure in prison. You can be food secure, in other words, and never have a say about what it is that you're provided, nor the manner in which it comes to you.

This predicament is also seen in the discrepancy between the qualitative and quantitative data that we have collected in this research. Even if a farmworker is food secure, according to the HFSSM, he or she is systemically deprived of the dignity to go to the grocery store independently or the autonomy to select the foods they wish to consume.

In contrast to food security, food sovereignty places the question of power at the forefront. Food security and food sovereignty are not necessarily points on a continuum, but rather are different approaches to understanding food access. Food sovereignty has been suggested as a decolonizing and more holistic approach to understanding food access, production, and distribution. *La Via Campesina* (the International Peasants' Movement) defines food sovereignty as,

> The right of people to healthy and culturally appropriate food produced through ecologically sound and sustainable methods and their right to define their own food and agricultural systems. It puts those who produce, distribute, and consume food at the heart of the food systems and policies rather than the demands of markets and corporations.
>
> (*Via Campesina*, Declaration of Nyéléni)

Food sovereignty takes many forms, scales, and approaches, but tends to highlight the ways by which neoliberalism and free-trade policies undermine and displace small farmers. The framework of food sovereignty intentionally positions itself in opposition to neoliberalism, structural racism, violence, and the legacy of colonialism, all factors that impact the realities of Latino/a farmworkers in the United States. The line between food security and food sovereignty is distinguished by the question of power. In the remainder of this chapter, we examine these questions of power in more detail, in part through returning to the story of Tomás and his garden.

Food access strategies in the northern borderlands

Through our fieldwork, we have found that the border has a definitive impact on farmworker food access. But we were also consistently surprised by the creative and varied ways in which farmworkers procure food. The following sections illuminate the major avenues that farmworkers in Vermont follow to bring food into the home.

Sending grocery lists

Farm owners, who also live in fear of the border patrol, encourage farmworkers to avoid public spaces, and they commonly purchase food for their employees. Typically, at least one farmworker in a household is literate in Spanish and using translational aids will write out a grocery list. Using this list, a farm owner or manager will buy groceries every 15 days. A survey by Naomi Wolcott-MacCausland (2010) found that in border counties 96.2 percent of migrant farmworkers rely on someone else to go to the store for them. Challenges with this system include barriers in language and literacy, a lack of knowledge about what foods are available, and limited choice in fresh, culturally specific foods. With the support of Northeast Sustainable Agriculture Research and Education (NESARE), Jessie Mazar developed and printed a bilingual visual grocery guide that is now being distributed by University of Vermont Extension personnel. This tool seeks to provide increased agency over food choices for farmworkers who are relying on a third party to go to the store and purchase food for them.

Mobile **tiendas** *(stores)*

One of the first times that we visited Lourdes, the wife of a farmworker from Oaxaca, she offered us *chicharones* (fried pork skins, a popular Mexican snack) and Valentina (a particular brand of hot sauce from Mexico). We were surprised to see this popular Mexican snack, given that these products are nearly impossible to find in Vermont's grocery stores. Over subsequent visits, we began noticing more such products—*guajillo* chili peppers, or *pan dulce* (sweet bread)—and started asking about their origins. Since then, we have learned about third-party food deliverers, who visit dairy farms on a regular basis to sell Mexican food products to the farmworkers. These food vendors bring products from Boston, New York, and New Jersey, sometimes at very elevated prices. In May 2016, while we were visiting Lourdes and her new baby, one of the food vendors came by in a trailer. Excited to see the operation, we surveyed the offerings as she looked through the shelves stocked with dried goods and familiar brands like Bimbo and Sabritas. The trailer was discreet from the outside, but on the inside it was a sanctuary of scarce Mexican food products. These mobile *tiendas* fill a major gap in access to culturally familiar foods. According to our interviews, farmworkers expect the mobile *tiendas* to arrive every 15 days. In some cases, the farmworkers will make a phone call to ask for a delivery, and other times the trailers will show up unannounced. Some of the vendors are from the United States but others are Latino/a immigrants.

Government programs

Another food access strategy for Vermont farmworkers is US-government support. Farmworker families who have given birth to children in the United States can utilize a small number of federally funded food access programs.

The two programs that interviewees reported using were Women, Infants, Children (WIC) and the free school-lunch program. In our interviews, people mentioned that the milk, eggs, beans, peanut butter, and cereal they received was useful, but they were not able to choose what food was delivered to them. Our research overlapped with a transition period in the WIC program from at-home delivery to a system whereby recipients would use an EBT (electronic benefits transfer) card to purchase approved foods from a grocery store, with Vermont being the last state to make this transition. The benefit of the EBT system is that it allows for more choice. But for migrant mothers with limited access to transportation, this transition presents additional obstacles in accessing these benefits.

In a stakeholder interview in 2015, a WIC employee in a border county said that the program was continuing home deliveries with farmworkers in the community for the time being because they did not have transportation and were often afraid of visiting the WIC office, which was located in a nearby town. The mothers qualify for WIC because their children are US citizens, but the parents are still criminalized for being here. In this same stakeholder interview, the WIC employee also said that the border patrol has followed her on home deliveries, parked behind her, and taken down her license plate information. While WIC and the school-lunch programs are important forms of support for farmworker families, accessing these benefits is not always a simple matter.

Huertas *gardening project*

In our research, the concept of fieldwork has taken a more literal meaning, as the majority of our work has been connected with our role in planning and planting gardens through Huertas. This gardening project germinated in response to the variety of pressures on undocumented farmworkers in northern Vermont: the pressure to be invisible, the fear of the border patrol, and the lack of mobility. The gardens offer a small reprieve where migrant farmworkers can access fresh vegetables and herbs that remind them of home. By connecting farmworkers with volunteers and materials and helping them obtain permission from the dairy owners to plant these gardens, Huertas aims to address the disparities in access to nutritious food, while simultaneously bridging the barriers of isolation and social inequalities.

Both of us became involved with the Huertas project in spring 2012, two years after our colleague Naomi Wolcott-MacCausland, a migrant health promoter with UVM Extension, recognized that there was a need for fresh, culturally familiar foods within farmworker households. In 2012, we formalized the program and planted gardens on 21 farms, reaching 82 adults and children. Over the next few years, the program continued, and in the 2017 season Huertas planted gardens on 27 farms, reaching approximately 80 to 90 adults and children. The program distributed 1,800 seed packets and plant starts, and with generous support from local growers and seed companies, now offers a selection of 94 varieties of vegetables, herbs, and flowers, including 11 varieties of chilies and culturally familiar herbs such as *epazote, papalo, hierbabuena,* rue, lemongrass, *chepiche*, chamomile, and aloe plants.

Tomás's garden

In 2014, when asked what he liked about Vermont, Tomás explained:

> Well, here my favorite is my work, no more than this. And beyond that, I like to plant tomatoes, chilies, radish, squash, this is not to sell products, it is just for the home. … I like when people come to visit, to feed them. It is not a business. This food, it is a gift of the earth, for everyone. It is not meant to be a business.

Here, Tomás eloquently captures the paradoxical nature of Huertas within the context and constraints of the US industrial food system. These qualities of reciprocity, generosity, community, and stewardship are fundamental to the reasons why Tomás enjoyed planting a garden. His personal philosophy on food was that it should always be shared. Coming from a small village in Mexico, Tomás had a vast amount of knowledge about food production, and he was infamous for filling up his visitors' cars with produce from his garden. There is a clear discrepancy between his personal philosophy on food being a gift of the earth and the rationalizing of the US industrial food system that has been his source of livelihood for most of his life.

The current political economic structure of neoliberalism values competition, efficiency, and consumption—with the principal metric of success being measured in profit. In many ways, the understanding of food security that the HFSSM measures represents this neoliberal emphasis on metrics, as opposed to a more holistic or emic understanding of food security. In contrast, food sovereignty centers a critique of neoliberalism and how it pervades our contemporary food systems. As suggested by Karl Polanyi, the framework of capitalism (the foundation for neoliberalism) values the human characteristic of competition and devalues the human characteristics of collaboration, redistribution, and reciprocity (McMichael 1996). Neoliberalism also devalues generosity, culture, community, and stewardship, unless they are reframed as commodities. In this system, where so many components of humanity are devalued, there is a tremendous amount of vulnerability. This explanation is helpful in reflecting upon the discrepancy between the philosophy of Tomás and that which underlies the US industrial food system. The economic systems in rural land-based communities value reciprocity and generosity as a way of allocating resources. Despite his role in sustaining the industrial food system, Tomás carries with him those same values as well as hospitality, all of which are inseparable from his emic understanding of the food that sustains him and those he cares about.

Over our the years of knowing Tomás, he often told stories of his time living and working in different parts of the United States—including tales of harvesting lettuce and running from police dogs in Arizona's fields, working in restaurants in Los Angeles, traveling to work in the back of 18-wheeler refrigerated trucks, and harvesting sugar beets in Idaho. At 62 years old, Tomás had an extensive résumé, having worked in the food system for more than 40 years of circular migration. He had lived in Vermont for 12 years when we interviewed him, and on that specific dairy for 7 years. A few months after our interview, he was interrogated by the border patrol after his housemate was spotted going to the grocery store with their employer. He was home resting between shifts when the border patrol knocked on the door. This is one of the greatest fears expressed by many farmworkers in Vermont.

Tomás spent a year waiting for a court date in Boston, where he was granted a delay of deportation for one year. He would joke and say that he could go to the store now because he had his papers—the court summons. He and his housemates were (somewhat ironically) able to move through public space with a bit more confidence after being detained. During this time, he planted a garden, worked on the farm, and prepared elaborate meals of *comida Mexicana*, which he enjoyed with everyone who visited his home. After a year of deliberation about the future of their farm, Tomás's employers ultimately decided to sell the herd for financial reasons. Tomás is now home, in Mexico.

Tomás's departure was a loss for the Huerta*s* community. He was incomprehensibly optimistic and dedicated to his work, garden, and community despite the lifelong marginalization he had faced working without documentation. But even in his departure, there was a lesson to be gleaned. It taught us that Huertas must remain a flexible project. Migrant farmworkers are necessarily mobile, while a garden is rooted in one place. Some Huertas participants have been on the same farm for many years, others for only a few months. After Tomás left, his housemate, Ernesto, became the caretaker of the garden space. When Tomás was on the farm, Ernesto didn't do much to maintain the garden, and we assumed that he was not interested. But Ernesto now cultivates an abundant garden, in the same space that Tomás cultivated for years.

Our research investigating food access in the northern borderlands has revealed that not only do farmworkers find creative ways to access culturally relevant food that satisfies caloric needs, but they also seek food that satisfies a deeper human need for cultural nourishment. Defining and measuring food security in narrow quantitative terms has minimal capacity to capture these complexities. Our research conducting HFSSM surveys was only able to capture a sliver of the lived experience related to food access in Vermont. The qualitative research allowed depth and understanding that illuminated the gaps we were able to identify in the survey data. Our understanding of food (in)security has been further complicated by each interview or garden visit, as the stories we heard were impossible to represent in the statistics generated from the food security surveys.

Conclusion

In conducting food access research in the northern borderlands, we have seen that the survey instrument to measure food insecurity simplifies the lived experience for farmworkers relating to food and does not do justice to capturing the reality of complex barriers and obstacles they face. Tomás's narrative captures a reoccurring theme among farmworkers in the northern regions of Vermont. This is an image of cultural perseverance and nostalgia through food, of the importance of family, and of the endurance of generosity and reciprocity. These glimmers of food sovereignty are seen in the tenacity and longing of these individuals to maintain their own cultural food, even while working within the US industrial food system. For Tomás, preparing a meal of *comida Mexicana* is a way to develop community and express his belief that "food is a gift of the earth." Meanwhile, his narrative also demonstrates the negotiations, barriers, and challenges that migrants face every day. Some of these barriers are physical, resulting in isolation and the inability to be out in the community, but there are other barriers that are more subtle. Negotiating these barriers requires compromise, including the difficult choices having to do with personal well-being and cultural identity.

Tomás's vegetable garden is perhaps an unassuming space whereby to consider notions of food sovereignty, yet it presents a clear vignette of resiliency and creativity in the face of structural oppression and vulnerability. Undocumented migrant farmworkers in the northern borderlands are made vulnerable by a political economic system that benefits from their illegality and criminalization. A consequence of this illegality is the contradiction that those who are working in Vermont's food system are experiencing significant challenges in accessing food themselves. But, as this chapter demonstrates, painting migrant farmworkers in Vermont as exclusively victims of a repressive system would present an incomplete picture. The thawing Vermont soil where gardens are planted in early June represents a space where food sovereignty can emerge, despite the fear of being stopped by the border patrol. Food sovereignty is realized among farmworkers in the amalgamation of these individually small acts, which add up and lead to reclaiming the notion of power, maintaining identity and culture, and creating opportunity where it seems there is none. This sovereignty is seen even in the smallest *epazote* seed that optimistically crosses the border to grow and flourish in a new land with the help of a migrant gardener.

References

Andreas, Peter. 2005. "The Mexicanization of the US–Canada Border: Asymmetric Interdependence in a Changing Security Context." *International Journal* 60(2): 449–462.

Anzaldúa, Gloria. 1987. *Borderlands/La Frontera: The New Mestiza*. San Francisco: Aunt Lute Books.

Borre, Kristen, Luke Ertle, and Mariaelisa Graff. 2010. "Working to Eat: Vulnerability, Food Insecurity, and Obesity Among Migrant and Seasonal Farmworker Families." *American Journal of Industrial Medicine* 53: 443–462.

Brown, Sandy, and Christy Getz. 2011. "Farmworker Food Insecurity and the Production of Hunger in California." In *Cultivating Food Justice: Race, Class, and Sustainability*, pp. 121–146. Cambridge, MA: MIT Press.

Carney, Megan. 2015. *The Unending Hunger: Tracing Women and Food Insecurity Across Borders*. Berkeley, CA: UC Press.

Cason, Katherine, Sergio Nieto-Montenegro, America Chavez-Martinez, Nan Ly, and Anastasia Snyder. 2004. "Dietary Intake and Food Security Among Migrant Farm Workers in Pennsylvania." Harris School of Public Policy Studies, University of Chicago, Working Papers.

Coates, Jennifer, Patrick Webb, and Robert Houser. 2003. "Measuring Food Insecurity: Going Beyond Indicators of Income and Anthropometry." Food and Nutrition Technical Assistance Project, Academy for Educational Development.

Coleman-Jensen, Alisha, Matthew P. Rabbitt, Christian A. Gregory, and Anita Singh. 2017, September. "*Household Food Security in the United States in 2016*." ERR-237. US Department of Agriculture, Economic Research Service.

De León, Jason. 2015. *The Land of Open Graves: Living and Dying on the Migrant Trail.* Oakland: University of California Press.

Declaration of Nyéléni. 2007. Retrieved from https://nyeleni.org/spip.php?article290.

Dhokarh, R., Himmelgreen, D. A., Peng, Y. K., Segura-Pérez, S., Hromi-Fiedler, A., & Pérez-Escamilla, R. 2011. "Food Insecurity is Associated with Acculturation and Social Networks in Puerto Rican Households." *Journal of Nutrition Education and Behavior.* 43(4): 288–294.

Essa, Jumanah S. 2001. "Nutrition, Health, and Food Security Practices, Concerns, and Perceived Barriers of Latino Farm/Industry Workers in Virginia." Master's thesis, Virginia Polytechnic Institute and State University.

FAO. 2016. "*Methods for Estimating Comparable Rates of Food Insecurity Experienced by Adults Throughout the World.*" Rome: FAO. Retrieved August 2018, from www.fao.org/3/a-i4830e.pdf.

Harrison, Gail G., Ame Stormer, Dena R. Herman, and Donna M. Winham. 2003. "Development of a Spanish-Language Version of the US Household Food Security Survey Module." *The Journal of Nutrition* 133(4): 1192–1197.

Himmelgreen, David, et al. 2007. "'I Don't Make the Soups Anymore': Pre-to Post Migration Dietary and Lifestyle Changes Among Latinos Living in West-Central Florida." *Ecology of Food and Nutrition.* 46(5): 427–444.

Holt-Giménez, Eric. 2014. "Vermont's Invisible Farmworkers." *HuffPost.* Retrieved February 2, 2018. www.huffingtonpost.com/eric-holt-gimenez/vermonts-invisible-farmwo_b_5518502.html.

Holmes, Seth. 2013. *Fresh Fruit, Broken Bodies: Migrant Farmworkers in the United States.* Oakland: University of California Press.

Kilanowski, Jill F., and Laura C. Moore. 2010. "Food Security and Dietary Intake in Midwest Migrant Farmworker Children." *Journal of Pediatric Nursing* 25: 360–366.

Kilanowski, Jill F., and Li Lin. 2013. "Effects of a Healthy Eating Intervention on Latina Migrant Farmworker Mothers." *Family & Community Health* 36(4): 350–362.

Kresge, Lisa, and Chelsea Eastman. 2010, June. "Increasing Food Security Among Agricultural Workers in California's Salinas Valley." *California Institute for Rural Studies*, www.cirsinc.org.

Mares, Teresa. 2019. *Life on the Other Border: Farmworkers and Food Justice in Vermont.* Oakland: University of California Press.

Mares, Teresa, Naomi Wolcott-MacCausland, and Jessie Mazar. 2013. "Cultivating Food Sovereignty Where There are Few Choices." Conference Proceedings from Yale University: *Food Sovereignty: A Critical Dialogue.*

McCandless, Susannah. 2010. *Conserving the Landscapes of Vermont: Shifting Terms of Access and Visibility.* PhD dissertation, Clark University, Worcester, MA.

McMichael, Philip. 1996. *Development and Social Change: A Global Analysis.* Thousand Oaks, CA: Sage.

Minkoff-Zern, Laura-Anne. 20102. *Migrations of Hunger and Knowledge: Food Insecurity and California's Indigenous Farmworkers*. PhD dissertation, University of California, Berkeley.

Moos, Katherine. 2008. *Documenting Vulnerability: Food Insecurity among Indigenous Mexican Migrants in California's Central Valley*. Washington, DC: Congressional Hunger Center.

Quandt, Sara A., Thomas A. Arcury, Julie Early, Janeth Tapia, and Jessie D. Davis. 2004. "Household Food Security Among Migrant and Seasonal Latino Farmworkers in North Carolina." *Public Health Reports* 119(6): 568–576.

Ready, Elspeth. "Challenges in the Assessment of Inuit Food Security." 2016. *Arctic* 69(3): 266–280.

Sano, Yoshie, Steven Garasky, Kimberly A. Greder, Christine C. Cook, and Dawn E. Browder. 2011. "Understanding Food Insecurity among Latino Immigrant Families in Rural America." *Journal of Family and Economic Issues* 32(1): 111–123.

Stoneman, Scott. 2010. "Raj Patel, on Rights, Sovereignty, and Suicide." Interview with Raj Patel. *Politics and Culture*, 2. Retrieved February 2018 from https://politicsandculture.org/2010/10/27/an-interview-with-raj-patel/.

Wadsworth, Gail, Thea Rittenhouse, and Sarah Cain. 2016. "Assessing and Addressing Farm Worker Food Security." *California Institute for Rural Studies*. Retrieved August 26, 2018, from www.cirsinc.org/phocadownload/assessing%20and%20addressing%20farm%20worker%20food%20security.pdf.

Weigel, Margaret M., Rodrigo X. Armijos, Yolanda P. Hall, Yolanda Ramirez, and Rubi Orozco. 2007. "The Household Food Insecurity and Health Outcomes of U.S.-Mexico Border Migrant and Seasonal Farmworkers." *Journal of Immigrant and Minority Health* 9(3): 157–169.

Wirth, Cathy, Ron Strochlic, and Christy Getz. 2007. "Hunger in the Fields: Food Insecurity Among Farmworkers in Fresno County, Davis, CA." *California Institute for Rural Studies*, 121–146.

Wolcott-MacCausland, Naomi. 2010. "Food Access Strategies in Border Counties." *University of Vermont Extension*. Retrieved through personal communication, 2010. Powerpoint slide.

4 Food insecurity and the struggle for food sovereignty in the time of structural adjustment

The case of Greece

Charalampos Konstantinidis

Introduction

Ten years after the global financial crisis of 2008, Greece still found itself in the most prolonged crisis in its history—and the deepest crisis in history for a European Union (EU) country. As the situation in Greece was primarily identified as a public-debt crisis, successive Greek governments and their lenders—the European Central Bank (ECB), the European Commission (EC), and the International Monetary Fund (IMF)—sought to address the crisis through a series of "structural adjustment programs" in 2010, 2012, and 2015. These programs offered Greece loans in exchange for a number of neoliberal "reforms": public sector and government spending reductions (austerity), wage decreases, privatizations, and legislation fostering a "business-friendly environment," as well as service and product market liberalization. Implementing such reforms and decreasing labor costs would purportedly increase the competitiveness of the Greek economy through "internal devaluation," attracting investment, restoring economic growth, and ensuring the long-term sustainability of the country's public debt (EC 2010).

Contrary to their promises, the austerity programs plunged the country into a major crisis. Gross domestic product (GDP) fell by approximately 25 percent compared to its pre-crisis level. Median income fell by more than 37 percent between 2010 and 2015, while incomes for lower-income groups fell by even more. The unemployment rate went from 12 percent before the signing of the first austerity program in 2010 to 28 percent by 2013.[1] Moreover, the economic crisis turned into a humanitarian crisis. Poverty and homelessness increased, and Greek households faced a significant decrease in their ability to make ends meet and cover their most pressing needs. By 2012, more than one in two households below the poverty line were considered food insecure, based on official statistics drawn from Eurostat's Statistics on Income and Living Conditions (EU-SILC) database. Furthermore, 60 percent of schoolchildren in certain districts were estimated to be food insecure and 25 percent of schoolchildren to be experiencing hunger (Deloitte 2014).

While the Greek experience constitutes a spectacular failure to meet the food needs of large segments of the population in a developed country, it also sparked responses seeking to mitigate food insecurity and transform Greece's agriculture and food systems. Such responses are connected to the global "food sovereignty" movement, which seeks to guarantee people's right to "healthy and culturally appropriate food produced through ecologically sound and sustainable methods, and their right to define their own food and agriculture systems" (La Via Campesina 1996 and 2007; see also Patel 2007; Magdoff and Tokar 2010). I argue that—despite promising efforts to transform the agriculture and food systems in Greece—the structural adjustment programs have created an uphill battle for food activists: the adjustment programs have not only extended and accelerated the neoliberal reform of the Greek agriculture and food systems but also restructured the governance of the Greek economy and society. As such, the structural adjustment programs challenge a key tenet of food sovereignty, namely the notion of respect for the "right of each nation to maintain and develop its own capacity to produce its basic foods respecting cultural and productive diversity" (La Via Campesina 1996).

I have organized this chapter in four sections, followed by a conclusion. First, I trace the rise in food insecurity during the crisis. I use Eurostat data to show that more than one in eight households face extreme levels of food insecurity. Second, I present grassroots responses to transform food production and distribution in crisis-ridden Greece. Third, I show how the three structural adjustment programs have been advancing neoliberal policies that achieve the further dismantling of domestic agricultural production and accelerate the centralization of the food sector. Fourth, I trace how the Third Structural Adjustment Program, in particular, demands the effective forfeit of sovereignty by the Greek people, thus undermining efforts to transform the agriculture and food systems in the direction of food sovereignty.

Food insecurity in the Greek crisis

In spite of increased awareness of the phenomenon of food insecurity after the global financial crisis and food riots across the world in 2007–2008, measuring food insecurity and developing a common understanding of food insecurity remains a challenge. To address this shortcoming, the Food and Agriculture Organization (FAO) launched in 2014 the Food Insecurity Experience Scale Survey Module, which poses eight questions about how lack of money and resources affect the food experience across different countries. These questions include whether the household was worried about not having enough food; was not able to eat healthy and nutritious food; ate only a few kinds of foods; had to skip meals or eat less than they thought they should; ran out of food; were hungry; or went without eating for a whole day (FAO 2016: 7). Based on this measure, the FAO estimates the prevalence of moderate and severe food insecurity in Greece in 2015 at 15.5 percent of the population. This was the highest value among all EU countries for which data is available, as well as substantially higher than the European average (8.8%).[2]

In EU official statistics, the only source for information on food insecurity is the Statistics on Income and Living Conditions (EU-SILC) database. This database is based on an annual survey conducted in all EU countries, which—among other social dimensions– calculates the share of households that report the "inability to afford a meal with meat, chicken or fish (or its vegetarian equivalent) every second day."[3] Inability to afford a full meal every second day is clearly a measure that only captures extreme instances of food insecurity (as compared to the FAO measure). As such, this measure has clear limitations in identifying more prevalent forms of food insecurity (such as skipping meals or resorting to low-quality food items). Nevertheless, this measure of material deprivation allows us to trace the evolution of food insecurity over time and to examine the effect of the crisis on food insecurity. Table 4.1 illustrates a striking rise in food insecurity in Greece after 2008: food insecurity rose from approximately 7 percent in 2008 to approximately 14 percent by 2012 and remains fairly stable at 13 percent since. For poorer households—whose income is below 60 percent of the country's median—this measure of food insecurity increased sharply after 2011 and exceeded half of these households (54%) by 2015.

Previous research has shown that even less extreme forms of food insecurity are associated with other measures of household stress, including financial (Gundersen, Kreider, and Pepper 2011); stress in meeting heating needs (the "heat or eat" dilemma) (Bhattacharya, DeLeire, and Haider et al. 2003); and stress in meeting medical needs (the "treat or eat" dilemma) (Berkowitz, Seligman, and Choudhry 2014). Since the EU-SILC measure of food insecurity captures only extreme instances of food insecurity, I present additional measures capturing these dimensions in Table 4.1. Note the sharp increases in the share of households that report only being able to make ends meet with difficulty or with great difficulty (from 54.5% in 2008 to 77.7% in 2015); in the share of households that are unable to face unexpected financial expenses (from 26.6% in 2008 to 53.4% in 2015); in the share of households unable to keep their house adequately warm (from 15.4% in 2008 to 29.2% in 2015); and the share of households with unmet medical needs due to cost (from 4.2% in 2008 to 10.9% in 2015). Thus, the increased prevalence of all these aspects of hardship indicates a potential increase in food insecurity during the crisis, beyond that of extreme food insecurity.

As I argue elsewhere in detail (Konstantinidis 2016), the rise in food insecurity during the crisis should be interpreted with a longer-term lens, and its underlying causes traced back to the restructuring of the agriculture and food systems in Greece since the country's entry into the European Economic Community (EEC) in 1981. This restructuring is, to some extent, an illustration of the uneven economic development of Europe and the diverging roles of "core" and "periphery" countries in the European Union—first, within the context of neoliberal policies and the completion of a common European

Table 4.1 Social indicators in Greece: 2003–2015

YEAR	*Inability to afford a meal with meat, chicken, fish (or vegetarian equivalent) every second day –% of population*		*Make ends meet with difficulty or great difficulty - % of population*	*Inability to face unexpected financial expenses - % of population*	*Inability to keep house adequately warm - % of population*	*Unmet medical needs – Reason: Too expensive – - % of population*
	TOTAL	*Below 60% of median equivalised income*	*TOTAL*	*Total*	*Total*	*Total*
2003	12.1	27.8	48	44.8	17.4	3.5
2004	8.3	20.5	45.4	35.5	16.8	3.1
2005	5.8	16.2	50.8	38.8	15.7	3.2
2006	7.9	21.3	52.6	30.6	12.0	4.5
2007	6.5	26.8	52.4	29.6	13.8	4.6
2008	7.1	29.9	54.5	26.6	15.4	4.2
2009	7.6	24.0	57	26.6	15.7	4.0
2010	7.9	22.7	58.4	28.2	15.4	4.2
2011	9.2	42.2	62.6	34.4	18.6	6.2
2012	14.2	50.7	73.1	40.5	26.1	6.5
2013	13.8	42.9	78.3	47.1	29.5	7.8
2014	13.0	47.5	78.1	51.8	32.9	9.7
2015	12.9	54.0	77.7	53.4	29.2	10.9

Source: Eurostat database, author calculations. Variables ilc_mdes03, ilc_mdes09, ilc_mdes04, ilc_mdes01, hlth_silc_21. Accessed February 3, 2017.

market, and accelerated after the creation of the Eurozone (Lapavitsas 2012; Vlachou 2012; Konstantinidis and Vlachou 2017).

This restructuring is characterized by two main elements. First, the dismantling of domestic production—illustrated by declining food production, stagnant agricultural productivity, and a rising food deficit—turned Greece into a food-importing country, thus delinking domestic labor costs and incomes from food prices. Second, the centralization of the food sector—exemplified by the increased presence of supermarkets and the monopolization of the food sector since the late 1980s—increased the power of food merchants over both farmers and consumers. These changes in the agriculture and food systems exacerbated lingering vulnerabilities that only became visible during the crisis. While incomes collapsed as an outcome of austerity and internal devaluation, food prices did not fall, because (a) food imports captured an increased share of food sales, and (b) food was sold in highly

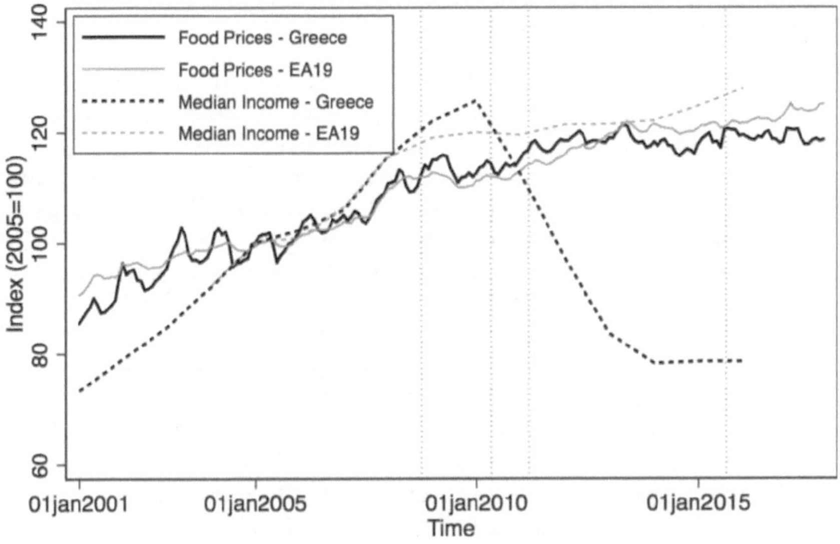

Figure 4.1 Food prices and median income: Greece and the Eurozone, 2001–2017.

Source: Eurostat, author calculations. Both food prices and median incomes are normalized (2005 = 100). The dotted lines delineate the beginning of the crisis (2008), and the three structural adjustment programs (May 2010, February 2012 and August 2015).

monopolized markets. This divergence between collapsing incomes and stagnant food prices was a major component of the struggle of Greek households to make ends meet and to cover their food needs.

Figure 4.1 presents the evolution of food prices and median income in Greece and the Eurozone. Before the crisis, as we see in Figure 4.1, food prices moved in tandem with the Eurozone average. Incomes in Greece, on the other hand, increased faster than the Eurozone average. The First Structural Adjustment Program, in 2010, marks the beginning of the strategy of internal devaluation, leading to a fall in median income by more than 37 percent between 2010 and 2015. For a significant part of the crisis (until 2013), however, food prices in Greece kept rising—and at even faster rates than prices in the Eurozone—despite the sharp fall in domestic incomes and labor costs. This is not surprising: increases in value-added taxes (VAT) on food as part of the effort to increase tax revenue during the crisis (EC 2010),[4] heavy dependence on imported food, and the role of middlemen and supermarkets explain the discrepancy between food prices and incomes. Thus, falling labor costs did not translate into lower prices for consumers, thereby compounding Greek people's difficulty to meet their needs.

Rebuilding the production and distribution of food in crisis-ridden Greece and food sovereignty

Against this context, a series of movements and initiatives in Greece sought to rebuild agricultural production and regain control over the distribution of food, to reconnect farmers with final consumers, and to challenge the effectiveness of existing institutions of food provisioning.[5] These movements shared common elements with the global movement for "food sovereignty" although they do not use that term. Food sovereignty is a term used by the peasant and food-activist movement seeking to resist a "neoliberal, industrial, capital-intensive and corporate-led model of agriculture" (Wittman, Desmarais, and Wiebe 2010) and to transform their food systems (both production and distribution) in ways that will secure both people's right to food and the protection of the ecosystem (La Via Campesina 2007; Patel 2009; Magdoff and Tokar 2010; Wittman et al. 2010). Thus, as ways to achieve people's right to food, food sovereignty calls for shifting away from monoculture and toward agroecological production methods, prioritizing small-scale agricultural production and local circuits of production and consumption, free access to seeds, supply-management policies, as well as protection against cheap imports (Rosset 2008; Patel 2009).

In Greece, the most significant phenomenon with respect to the reorganization of *production*—which in general has remained at infant stages—was the emergence of urban gardens. The first such instances were guerilla gardens, associated with conflict over land use in dense urban settings—such as the Navarino Park in the Exarchia neighborhood in Athens, which was converted from a parking lot into a self-managed park.[6] After 2011, urban gardens were "mainstreamed," with several municipal authorities throughout Greece offering garden allotments as part of the attempt to alleviate poverty among their residents (Anthopoulou and Nikolaidou 2013; Partalidou and Anthopoulou 2015).

Changes in the *distribution* of food have been more significant. In February 2012, the "no middlemen movement" (also known as the "potato movement") sprang up in northern Greece when a volunteer group from the city of Katerini contacted potato producers from the town of Nevrokopi, arranging a 24-ton sale of potatoes at a price of 0.25/kg, thus allowing farmers to dispose of unsold product and consumers to buy potatoes at one third of supermarket prices.[7] The success of this first initiative captured widespread attention and was replicated in other cities (Calvario and Kallis 2017). Such interactions have taken multiple forms: either occasional disposals of food at prearranged dates and times or, as in the city of Agrinio, markets exclusively for local farmers. Moreover, the first experiments with community-supported agriculture took place in Athens, connecting organic farmers from the metropolitan region with urban consumers.[8]

In addition to markets facilitating direct producer-consumer interactions, other types of food institutions have sprung up in Greece during the crisis. Food pantries and soup kitchens have now been set up in most cities, helping people meet their basic food needs while challenging the commodification of food. Moreover, collective/social kitchens aim not only to provide food, but also to involve and mobilize those in need, highlighting the collective character of food and breaking down the stigmatization and the power hierarchy associated with receiving food.[9] Such attempts try to break down ethnic and national barriers around food, creating inclusive food movements and offering a progressive vision around food, as well as countering the repeated attempts of the neo-Nazi Golden Dawn group to capitalize on poverty through its highly publicized handouts of food "to Greeks only."

Structural adjustment and the state

The sprouting of solidarity initiatives may appear to be a success of volunteerism, the social and solidarity economy, or third-sector initiatives. Nevertheless, the role of the state for their flourishing should not be ignored. As Parenti (2015) argues, there is a recent tendency within the Left to avoid analyzing the role of the state for the valorization of capital and the functioning of capitalism, and to assume that transformative movements can ignore questions of state power. This is particularly relevant to the flourishing of food initiatives described above. The state (at both central and local levels) can either support or limit the success and scope of solidaristic structures by changing the legal framework in which they operate. Moreover, the state can offer material support to such solidarity initiatives through either direct funding or other forms of material support. On the other hand, by adopting structural adjustment programs, the state can transfer power from its own citizens to its lenders, and create new conditions for the operation of capital.. All these changes—as well as their relation to the goal of "food sovereignty"—deserve closer scrutiny.

In the Greek case, the different levels of the state have taken contradictory positions during the crisis. In certain instances, particularly in towns with progressive mayors, local authorities have promoted solidarity structures by providing space and logistical support for producers' markets and "no-middlemen" initiatives. Moreover, municipal authorities in several cities have provided city-owned space for urban gardens or set up social pantries, often in collaboration with non-governmental organizations (Calvario and Kallis 2017).

At the central level, Greek governments have largely undermined efforts to rebuild the Greek agriculture and food systems. Most importantly, by pursuing the structural adjustment programs (EC 2010, 2012a, 2015), successive Greek central governments have implemented austerity and policies that expand capital's scope of operations in Greece. Similar to other groups, farmers faced increased taxation on their income, as well as reductions of tax privileges (such as diesel subsidies and VAT refunds) (EC 2012b; 2013: 29; 2015: 6–8). In combination with generalized increases in VAT and energy prices, farmers saw significant increases in non-labor input costs, making it more difficult for them to make ends meet in the context of decreased domestic consumption and harder access to lending.[10]

Moreover, the structural adjustment programs have aimed to further deregulate and liberalize agricultural services and product markets, aligning Greek legislation with European Union regulation and extending the completion of the European market. Perhaps the most striking instance of liberalization revolves around the creditors' demand that Greece abolish national legislation that limited the shelf life of fresh milk to five days (EC 2014; 2015). The explicit goal of this intervention—which was in line with the Organization for Economic Cooperation and Development (OECD) (2014) recommendation—was to open the Greek market to milk imports from other EU countries. According to the OECD, this change would increase competitive pressure on Greek farmers, making it hard for "inefficient small local farms" to operate unless they increased their productivity (OECD 2014: 55). In line with these recommendations, the Greek government increased the maximum shelf life of fresh milk to seven days in 2014 (EC 2014), a change that however yielded no consumer price reduction. This is hardly surprising: as the OECD itself recognizes, milk producer prices are only 35 percent of milk consumer prices in Greece (OECD 2014: 50), the difference between consumer and producer prices in Greece being 35 percent higher than the EU average. Such divergences indicate the presence of high markups in the intermediate stages of processing and food retail (OECD 2014: 51). In other words, if one wanted to reduce consumer prices, one would have to curtail the high market power of intermediate actors in the Greek food chain.

In fact, rather than taking measures to curtail such high markups, the structural adjustment programs spearheaded a series of changes that strengthened the position of large players and led to more centralization in the food retail sector. Ironically, a number of these changes—such as the introduction of Sunday trading in 2013—occurred in the name of product market competition and of lifting barriers to competition. At the same time, the results of such policies disproportionately benefit large retailers over small ones. Accordingly they contribute to an ongoing monopolization of the retail sector, exemplified in a wave of supermarket mergers and acquisitions, which has allowed the three largest retailers (Vasilopoulos, Sklavenitis, and Lidl) to increase their sales at a time when overall sales remain stagnant, thus consolidating their position during the crisis.[11] Furthermore, in May 2014, the central government changed the legal framework in line with OECD (2014) and lender recommendations to "liberalize outdoor trade" (EC 2014: 248). Thereby, the government banned the sale of products by outdoor traders in the vicinity of stores selling similar products and in municipalities of more than three thousand people, and effectively outlawed the "no middlemen" movement in the entire country. At the same time, by placing restrictions on outdoor trading in the vicinity of stores, the Greek government created a framework that protected the interests of retail stores—supermarkets in particular—which were challenged by the new types of direct producer–consumer interactions.

Food sovereignty without sovereignty?

The election of the Coalition of the Radical Left (Syriza) to lead the government in January 2015 offered an opportunity for mitigating food insecurity and rebuilding the agricultural system of Greece. While in opposition, Syriza had promised to end austerity and the humanitarian crisis in Greece—as well as to challenge capitalist hegemony in the economy through building alternative institutions that prioritize people's needs (see Tolios 2009; Laskos and Tsakalotos 2013: 142–143). Furthermore, Syriza had supported the "no middlemen" movement, as well as other solidarity structures. In fact, the first act of the Syriza government in March 2015, in defiance of the country's lenders, was passing a law to face the "humanitarian crisis." This law introduced a program of food stamps, providing monthly assistance for food purchases to approximately 145,000 low-income Greeks; free electricity to low-income families; and installment payment schemes for indebted lower-income households.[12]

Syriza's defiance of the country's lenders did not last long. In summer 2015, having miscalculated its leverage against its European creditors, and faced with the country's expulsion from the Eurozone, the Syriza government agreed to what it had previously denounced: a Third Structural Adjustment Program in exchange for new loans. Thereby, the Syriza government (which won a renewed mandate after the September 2015 elections) committed to undertake measures that the previous right-wing government had been unable to push through, such as additional tax increases, new privatizations, and an overhaul of the country's pension system. Furthermore, the Greek government committed to not introduce any new installment or settlement schemes for debt payments similar to the ones introduced through its law to face the 'humanitarian crisis' (EC 2015). Additionally, in October 2015, the Syriza government reversed several key provisions of the law to face the humanitarian crisis: in particular, the Greek government increased interest rates for participants in the installment scheme and revoked the installment option for late payers.

The Third Structural Adjustment Program had significant ramifications for Greek farmers. Greece committed to abolish the refund of excise tax on diesel oil farmers by October 2016. Moreover, the Greek government agreed to increase farmers' social contributions as well as their taxes—both in the form of higher tax rates and through increased prepayment of value-added taxes (EC 2015:6–7).

The Third Structural Adjustment Program sought to extend the liberalization of product markets. In the dairy sector, the Greek government committed to conform to the earlier OECD recommendation of letting producers set the use-by date of milk, altogether abolishing the shelf-life regulation. The Syriza government also committed to undertake the remaining OECD toolkit recommendations, inter alia, allowing supermarkets to sell over-the-counter drugs and to bake bread from frozen dough (OECD 2014; EC 2014, 2015). Furthermore, in 2017 the Syriza government reduced previous restrictions on Sunday trading for supermarkets (EC 2017: 18). All these measures led to the strengthening of the competitive position of large stores able to absorb additional labor and operating costs more easily than smaller stores and, inevitably, to consolidate corporate control over the food system at the expense of small farmers and consumers.

Ironically, in August 2015, at the same time that the Syriza government was committing to the Third Structural Adjustment Program, it asserted that it would counteract its recessionary effect through a "parallel program" (Syriza 2015: 10). Among other elements, this program called for a new agricultural and food model that would reverse the environmental crisis and help achieve food security through sustainable agriculture and the support of small farmers producing traditional crops (Syriza 2015: 19–20). The Syriza program sought to increase Greek agricultural production through the promotion of cooperative and solidarity initiatives; healthy, safe, nutritious and affordable food for consumers; social planning around food rather than complete control by market forces; and shortening the food chain between producers and consumers (Syriza 2015: 19–20). To this end, it suggested utilizing idle farmland, as well as revitalizing the Greek food and drink industry along social economy lines (Syriza 2015: 25). Moreover, it called for the introduction of school meals for elementary school pupils, the promotion of new legislation around food-waste management, and the legalization of the "no-middlemen" movement (Syriza 2015: 39). Although the document does not use the term, one could plausibly claim that its aims are aligned with the principles and goals of food sovereignty. In fact, the "parallel program" underlines food as a primarily social and democratic issue, and explicitly characterizes rebuilding food production in Greece as a weapon against future blackmail.[13]

While these plans may appear to move to the direction of restoring agricultural production and creating new forms of food distribution, the Third Structural Adjustment Program significantly constrains Greece's ability to govern its food system. First, according to the program, the Greek government commits to "*consult and agree* with the European Commission, the European Central Bank and the International Monetary Fund on *all actions* relevant for the achievement of the objectives of the Structural Adjustment Programme of Understanding before these are finalized and legally adopted"

(EC 2015: 4, emphasis added). The requirement to *consult and agree* heightens lender supervision compared to the previous structural adjustment programs, which instead bound the Greek authorities to *consult* with its lenders before adopting legislation (EC 2010: 59, 2012a: 94).

Second, the Third Structural Adjustment Program calls for the depoliticization of Greek public administration in crucial sectors such as revenue administration (EC 2015: 5). This presumes that public administration can be reduced to technical tasks that are separated from political direction and do not change irrespective to changes in power. Furthermore, this "reform" attempts to build a state, crucial parts of which are run by technocrats freed from the pressure of democratic accountability—just like the neoliberal prescript of central bank independence has "freed" central bankers from pressure to pursue low-unemployment policies. Thus, the depoliticization of administration represents a veiled but serious challenge to democratic sovereignty.

Finally, a third way in which the Third Structural Adjustment Program constrains sovereignty is by stipulating that

> [n]o unilateral fiscal or other policy actions will be taken by the authorities, which would undermine the liquidity, solvency or future viability of the banks. All measures, legislative or otherwise, taken during the programme period, which may have an impact on banks' operations, solvency, liquidity, asset quality etc. should be taken in close consultation with the EC/ECB/IMF and where relevant the ESM.
>
> (EC 2015: 17)

It is obviously impossible to imagine legislative or policy measures that categorically have no impact on banks' operations or asset quality. In the current Greek context, in particular, this raises questions about the future structure of land ownership and the agricultural sector. Given high private-sector indebtedness, such stipulations could open up the potential for stripping land from indebted farmers who have used farmland as collateral. Finally, such a provision reiterates a key feature of neoliberal capitalism: the primacy of financial interests over competing demands, such as, for instance, demands to provide debt relief to small farmers.

Thus, through such stipulations, the Third Structural Adjustment Program goes beyond promoting specific policy measures and establishes unprecedented forms of governance for an EU member. By subjecting all proposed legislation in Greece to the approval of organizations that are not democratically accountable (such as the ECB and the IMF), the Third Structural Adjustment Program constrains public control over the direction of policy.

For food and agriculture, in particular, it constrains the extent to which the state can support changes towards food sovereignty, even if a governing party purportedly embraces this goal. The years of the Syriza administration illustrate the contradictions of trying to build food sovereignty without political sovereignty. On the one hand, Syriza has pushed through some progressive measures around food. Beyond introducing food stamps in 2015, in 2017 the government institutionalized "consumers' markets," a form of markets that seek to support small producers by bypassing intermediate actors (*Greek Government Gazette* 2017: 3382). Moreover, in 2017, the government expanded a pilot school meal program to include 130,000 elementary school pupils, covering approximately 20 percent of the elementary school student population.[14]

While these policies are progressive, they do not suffice to speak of progress toward food sovereignty. Continuing austerity extends the pauperization of large segments of the population of Greece, while liberalization seeks to further open up the Greek market to imports and to push out small farmers. Moreover, more ambitious policies such as those supporting rebuilding domestic production with active measures would be considered a reinstatement of protectionism—and a violation of the European single market—therefore likely to be struck down by the country's lenders, who have established themselves as the ultimate arbiters of formal governance.

Thus, Greece's new structure of governance highlights the limitations of what food movements can achieve when the state is unwilling (or unable) to pursue radical policies toward food sovereignty. This experience also illustrates that structural adjustment programs do more than simply impose austerity. More importantly, they seek to foreclose the building of radical alternatives, such as food sovereignty, by instituting new methods of governance that reshape control over territory and legislation. Although food movements are a significant component of the struggle for food sovereignty, they cannot legislate and implement policies that would allow the people to define and control their agriculture and food systems. Thus, considering the changing nature of state power in times of structural adjustment, and returning power from the hands of lenders to the people, becomes an essential part of the struggle for food sovereignty.

Conclusion

The preceding sections document the dramatic rise of extreme food insecurity in Greece during and after the global financial crisis of 2007–2008. I argue that the rise of food insecurity during this period has to be understood in the context of the restructuring of the agriculture and food systems of Greece prior to the crisis. The lingering vulnerabilities of the Greek agriculture and food systems were revealed during the crisis when the strategy of internal devaluation created a large divergence between falling incomes and stagnant food prices. Large segments of the Greek population experienced increased difficulty meeting their most basic needs, including food. But given the dependence on imports and the high degree of centralization in the food sector, food prices did not fall, exacerbating the pauperization of large parts of the Greek population.

The rise in food insecurity elicited a series of initiatives that sought to reshape the Greek agriculture and food systems. Efforts, in particular, to reconnect small producers and consumers, circumventing middlemen and supermarkets, share common characteristics with the global food sovereignty movement. But the Structural Adjustment Programs adopted by successive Greek governments created an uphill battle for farmers and food activists. These programs have extended the liberalization of the food sector—further dismantling domestic production—while also strengthening the role of supermarkets and their power over both farmers and consumers. Moreover, the Third Structural Adjustment Program seeks to establish new forms of governance that constrain the sovereignty of Greek people, as they establish lender veto-power over all proposed legislation and limit the Greek people's control over the state.

The Greek experience also poses important political and theoretical questions. The first question concerns the need for food activists to pay attention to economic policies seemingly unrelated to food and how they affect people's ability to meet their basic needs. As the Greek case illustrates, public debt has been used as a lever to execute a wide array of neoliberal reforms with far-reaching implications for people's access to food. The second question concerns the role of the state. Activists and progressive academics often highlight grassroots responses in mitigating food insecurity but tend to put little emphasis on the interactions between these efforts and the state. However, the effectiveness and viability of grassroots attempts to transform the agriculture and food systems cannot be completely separated from changes at the state level, especially in the context of radical changes to the exercise of power. Without ignoring the importance of movements, more attention needs to be placed on the state for countries to move toward food systems that promote ecological sustainability and allow people to meet their most basic needs.

Notes

1 Eurostat database series: nama_10_gdp (GDP); une_rt_m (unemployment); ilc_di01 (income distribution), retrieved December 1, 2016.
2 See https://unstats.un.org/sdgs/indicators/database/?indicator=2.1.2, retrieved February 23, 2018.
3 See http://appsso.eurostat.ec.europa.eu/nui/show.do?dataset=ilc_mdes03&lang=en, retrieved February 25, 2018.
4 The average VAT rate on food increased by 75% during the crisis (from 10.8% in 2008 to 18.9% in 2017). See IELKA 2017.
5 See www.avgi.gr/article/3318390/stin-koinonia-tis-krisis-i-allileggui-einai-pantou-, retrieved January 17, 2016.
6 See http://parkingparko.espivblogs.net/englishfrench/about-the-park/, retrieved January 17, 2016.

7 See www.agronews.gr/agora/organoseis/arthro/77893/to-kinima-tis-patatas-apo-tin-pieria-os-tin-kriti/, and www.aljazeera.com/indepth/features/2012/06/201261 1102126662269.html, retrieved January 18, 2016.

8 For a list of solidarity initiatives, see solidarity4all.gr and enallaktikos.gr, accessed January 17, 2016. For community-supported agriculture, see http://agronaftes.blogspot.com/p/csa.html and www.paseges.gr/el/news/Exaplwnetai-h-koinotika-yposthrizomenh-gewrgia, retrieved January 17, 2016.

9 See http://oallosanthropos.blogspot.co.uk/p/social-kitchen-other-human.html, retrieved January 30, 2017.

10 Fuel and electricity prices increased in Greece by 102% between 2009 and 2013, as opposed to 25% for all Eurozone countries, to a large extent due to increased taxation. See Eurostat database, series: prc_hicp_midx (consumer prices), retrieved December 17, 2015.

11 See. www.kathimerini.gr/939472/article/oikonomia/ellhnikh-oikonomia/e3afanis thkan-oi-mises-alysides-soyper-market-se-mia-oktaetia, www.kathimerini.gr/938 437/article/oikonomia/epixeirhseis/nees-isorropies-diamorfwnontai-pleon-sto-lianemporio-trofimwn, and www.kathimerini.gr/915036/article/oikonomia/epixeirh seis/se-av-kai-lidl-strafhkan-oi-pelates-ths-marinopoylos, retrieved February 25, 2018.

12 See www.bbc.com/news/world-europe-31952521, as well as www.anthropocrisis.gr/anthropocrisis/Pages/Info.aspx for specific benefits and eligibility criteria, retrieved January 21, 2016.

13 This refers to the idea that—given Greece's large food deficit—an exit from the euro would have led either to food shortages, or to a foreign exchange crisis.

14 See http://dipe.mes.sch.gr/autosch/joomla15/2017/December/01-01-sitisi-programma-09-11-2017.pdf, retrieved February 23, 2018.

References

Anthopoulou, Theodosia, and Sophia Nikolaidou. 2013. Δημοτικοί αστικοί λαχανόκηποι και βιώσιμη πόλη: τοποθετώντας το αγροδιατροφικό σύστημα στην αστική ατζέντα [Municipal urban food gardens and sustainable city: placing the agro-food system on the urban agenda]. [in Greek]. Retrieved February 25, 2018, from http://grsa.prd.uth.gr/conf2013/49_anthopoulou-nikolaidou_ersagr13.pdf.

Berkowitz, Seth. A., Hilary K. Seligman, and Niteesh K Choudhry. 2014. "Treat or Eat: Food Insecurity, Cost-Related Medication Underuse, and Unmet Needs." *American Journal of Medicine* 127(4): 303–310.

Bhattacharya, Jayanta, Thomas DeLeire, Steven Haider, and Steven and Janet Currie. 2003. "Heat or Eat? Cold-Weather Shocks and Nutrition in Poor American Families." *American Journal of Public Health* 93(7): 1149–1154.

Calvario, Rita, and Giorgos Kallis. 2017. "Alternative Food Economies and Transformative Politics in Times of Crisis: Insights from the Basque Country and Greece." *Antipode* 49(3): 597–616.

Deloitte. 2014. "Διατροφή" Program Evaluation. Stavros Niarchos Foundation. July 23.2014. Retrieved February 25, 2018, www.snf.org/media/4427555/SNF_Nutrition_Program_Assessment.pdf.

European Commission. 2010. "The Economic Adjustment Programme for Greece." Occasional Papers 61, May 2010. Brussels: Directorate—General for Economic and Financial Affairs.

European Commission. 2012a. "The Economic Adjustment Programme for Greece." Occasional Papers 94, March 2012. Brussels: Directorate—General for Economic and Financial Affairs.

European Commission. 2012b. "The Second Economic Adjustment Programme for Greece: First Review." Occasional Papers 123, December 2012. Brussels: Directorate—General for Economic and Financial Affairs.

European Commission. 2013. "The Second Economic Adjustment Programme for Greece. Second Review—May 2013." Occasional Papers 148, May 2013. Brussels: Directorate—General for Economic and Financial Affairs.

European Commission. 2014. "The Second Economic Adjustment Programme for Greece: Fourth Review." Occasional Papers 192, April 2014. Brussels: Directorate—General for Economic and Financial Affairs.

European Commission. 2015. "Memorandum of Understanding between the European Commission Acting on Behalf of the European Stability Mechanism and the Hellenic Republic and the Bank of Greece." Retrieved December 18, 2015, http://ec.europa.eu/economy_finance/assistance_eu_ms/greek_loan_facility/pdf/01_mou_20150811_en.pdf.

European Commission. 2017. "Compliance Report: ESM Stability Support Programme for Greece. Second Review—June 2017." Retrieved February 11, 2018, https://ec.europa.eu/info/sites/info/files/compliance_report-to_ewg_2017_06_21.pdf.

FAO. 2016. *Methods for estimating comparable rates of food insecurity experienced by adults throughout the world.* Rome, FAO. www.fao.org/3/a-i4830e.pdf.

Greek Government Gazette. 2017. "FEK (A-171). Law 4491/2017." November 13, 2017. Retrieved [in Greek] February 12, 2018, www.et.gr.

Gundersen, Craig, Brent Kreider, and John Pepper. "The Economics of Food Insecurity in the United States. *Applied Economic Perspectives and Policy* 33(3): 281–303.

IELKA. 2016. Η επίδραση των συνεχών αλλαγών ΦΠΑ στα είδη παντοπωλείου (τρόφιμα) τη δεκαετία 2008–2017. [The effect of constant VAT changes on food items during the 2008–2017 decade] [in Greek]. Retrieved February 10, 2018, www.ielka.gr/?p=2039.

Konstantinidis, Charalampos. 2016. "The Neoliberal Restructuring of Agriculture and Food in Greece." *Review of Radical Political Economics* 48(4): 544–552.

Konstantinidis, Charalampos, and Andriana Vlachou. "Appropriating Nature in Crisis-Ridden Greece: The Rationale of Capitalist Restructuring, Part 1." *Capitalism Nature Socialism* 28(4): 48–61.

Lapavitsas, Costas. 2012. *Crisis in the Eurozone.* London: Verso.

Laskos, Christos, and Euclid Tsakalotos. 2013. *Crucible of Resistance: Greece, the Eurozone & the World Economic Crisis*. London: Pluto Press.

Magdoff, Fred, and Brian Tokar. 2010. "Agriculture and Food in Crisis: An Overview. In Fred Magdoff and Brian Tokar (eds.). *Agriculture and Food in Crisis: Conflict, Resistance and Renewal* pp. 9–30. New York: Monthly Review Press.

OECD. 2014. *Competition Assessment Review: Greece.* Paris: OECD Publishing.

Parenti, Christian. 2015. "The 2013 ANTIPODE AAG Lecture: The Environment Making State: Territory, Nature, and Value." *Antipode* 47(4): 829–848.

Partalidou, Maria, and Theodosia Anthopoulou. 2015. "Urban Allotment Gardens During Precarious Times: From Motives to Lived Experiences." *Sociologia Ruralis* 57(2): 211–228.

Patel, Raj. 2007. *Stuffed and Starved: The Hidden Battle for the World Food System*. Brooklyn: Melville House Publishing.

Patel, Raj. 2009. "Food Sovereignty." *The Journal of Peasant Studies* 36(3): 663–706.

Patel, Rajeev, Radhika Balakjrishnan, and Lima Narayan. 2008. "Transgressing Rights: La Via Campesina's Call for Food Sovereignty, *Feminist Economics* 13(1): 87–116.

Rosset, Peter. 2008. "Food Sovereignty and the Contemporary Food Crisis." *Development* 51(4): 460–463.

Syriza. 2015. "Government Program Plan, August 2015" [in Greek]. Retrieved January 21, 2016, www.avgi.gr/documents/10179/0/ΣΧΕΔΙΟ%20ΠΡΟΓΡΑΜΜΑΤΟΣ%20ΣΥΡΙΖΑ/b29e5a3e-41b8-4a17-b7cc-caba9d0e68af.

Tolios, Y. 2009. Περιβάλλον και Αγροτική Πολιτική σε συνθήκες παγκοσμιοποίησης: Εναλλακτική Στρατηγική Αυτοδυναμίας Τροφίμων (Food Sovereignty) [Environment and Agricultural Policy during Globalization: An Alternative Strategy of Food Sovereignty. Athens: Kapsimi.

Via Campesina. 1996. The Right to Produce and Access to Land. Retrieved January 25, 2017, www.acordinternational.org/silo/files/decfoodsov1996.pdf.

Via Campesina. 2007. "Declaration of Nyeleni." Retrieved January 25, 2017, https://nyeleni.org/spip.php?article290.

Vlachou, Andriana. 2012. The Greek Economy in Turmoil. *Rethinking Marxism* 24(2): 171–200.

Wittman, Hannah, Annette Aurelie Desmarais, and Nettie Wiebe. 2010. "The Origins and Potential of Food Sovereignty." In Hannah Wittman, Annette Aurelie Desmarais, and Nettie Wiebe (eds.). *Food Sovereignty: Reconnecting Food, Nature and Community*, pp. 1–14. Black Point: Fernwood Publishing.

5 Food insecurity in the age of neoliberalism in Turkey and its neighbors

Murat Öztürk, Faik Gür, and Joost Jongerden

Introduction

The political economy of agriculture in the northern part of the Middle East has drastically changed over the last decades, during which a fundamental shift from a developmentalist model to neoliberal economic policies took place.[1] Since the 1980s, these policies have penetrated almost all aspects of economic and social life and thus have been strongly affecting agriculture and food systems in the region. In this chapter, we focus on the issue of food security in Turkey in the age of neoliberalism.

Turkey adopted neoliberal policies as early as the 1980s in all the sectors of its economy, agriculture being the latest (Pamuk 2014; Senses 2016). As the agriculture sector in Turkey eventually became export oriented and market based, it began to consider its neighbors as potential markets. Turkey's borders with Iraq and Syria are the longest (1,295 km, 384 and 911 km respectively), which allowed for relatively easier international road transport compared to other neighbors. Thus, agricultural products from Turkey began to pour into these two countries more than any other neighboring countries in the Middle East, which also explains why neoliberalism is a regional phenomenon as well as a global one. Because of the economic ties between Turkey and its neighbors to the south and east, it is important to present briefly the cases of Syria and Iraq in the context of the impact of neoliberal policies on food insecurity in Turkey.

Food security had been an objective for the countries in the region, yet neoliberal policies structurally undermined agricultural production. Although food security had been an objective of the government in Syria in the 1980s and 1990s, economic liberalization in the early 2000s in combination with food price spikes in 2007 and 2008, resulted in an increase in poverty and reduced economic access to food in the country, particularly in rural households (FAO 2001; Hindi 2011; Rocchi et al. 2013; Gobat and Kostial 2016). It is little wonder, therefore, that the failed neoliberal economic reforms in the agricultural sector in Syria are thought to have played a critical role in the emergence of the civil war there (Gleick 2014). As for Iraq, demographic de-ruralization and urban growth have negatively affected agricultural productivity and increased dependence on food imports since the mid-1960s (Schnepf

2004). The UN sanctions against the Saddam regime after 1991 severely affected households' physical access to food, which was addressed through a public distribution system based on cheap imports. The 2003 invasion and the neoliberal aftermath left agriculture in disarray throughout much of the country (Moore and Parker 2007).[2] The US-installed provisional authority and subsequent governments embarked on a program of neoliberalization. The agricultural sector was left without adequate government support and thus processing capacity collapsed, as did the infrastructure for the provision of extension (facilities and functioning), the procedures for food and product safety, and the applied research (Beer 2016: 295; Gibson 2012).

It is against this background of neoliberal policies and upheaval in the region that we here discuss food security issues in neighboring Turkey, whose agricultural structure, dominated by small-scale producers, has helped to secure its position as one of the self-sufficient, thus food secure, countries in the region.[3] But as Turkey completed a structural adjustment processes in the early 1990s (Aydın 2010; Öztürk 2012b; Pamuk 2014), one that included all the sectors of the national economy, a major problem developed in the area of food accessibility. The critical food studies literature began to address this issue in general during the neoliberal period (Bernstein et al. 1990; Friedmann and McMichael 1987; Koc, Sumner, and Winson 2017).[4]

Drawing on various national and international data sets, we argue in this chapter that food security is weaker in countries and regions where there is a combination of conflict (military unrest, civil war, etc.) and neoliberal agricultural policies, and that this holds regardless of how powerful their agriculture has been in the past—even in Mesopotamia and Anatolia (today's Syria, Iraq, and Turkey), the birthplace of agriculture. For analysis of the consumer side, we use IPSOS[5] household consumer panel data from the 1980s.This analysis is based on correlation with the income-level information in the IPSOS Turkey database. For the production side, we look at the cost of agricultural inputs and changing trends in agricultural incomes. An overview of official statistics on agricultural input and agricultural product prices shows that farmers in Turkey have been losing income in real terms during the period under investigation, indicating that production capabilities are under threat. We employ additional data to support our initial finding of diminishing real income, showing farmers to have fallen behind on credit payments and some to have eventually abandoned agricultural production. Finally, we use the results from a recently (2017) completed field study that provides insights into traditional homemade food production by the household for household use trends in villages across Turkey.

A troubled region

In Syria, food security concerns among the rural power base of the leading Ba'ath Party have informed support for the agricultural sector. Self-sufficiency in key staple foods, such as wheat, was obtained and poverty reduction realized (IFAD 2001), yet within the Ba'ath Party there was a push for economic liberalization. This began as early as the 1980s (Barnes 2009), although neoliberal economic restructuring was only fully implemented in the 2000s. Agreements were signed to accelerate trade liberalization (Al-Hindi 2011). In an attempt to shift growth to the urban sector, subsidies on basic goods were slashed (Lund 2014). The neoliberal restructuring changed the powerbase of the Ba'ath. It created a loyal business sector but moved away from its popular powerbase. Though being promoted by international institutions,

> the shift to a market economy ... failed to meet the basic goals of achieving higher living standards," so, "while a small group ... amassed unprecedented levels of wealth because of privatization and liberalization, a majority of public sector workers and peasants ... lost access to means of survival.
>
> (Joya 2010)

As a result, the liberalization of the agricultural sector after 2000 led to a significant decrease in agricultural employment. Some 460,000 people (33% of the total work force) discontinued work in the agricultural sector between 2001 and 2007, many migrating to cities where employment opportunities were relatively few, with living expenses high and purchasing power low, while austerity measures undermined social support systems (Abboud 2013; Chatel 2014). As a consequence of economic liberalization, a bourgeoisie emerged, often from those loyal to the Ba'ath Party, but this signaled a return to impoverishment for the rural population, as "poverty of the pre-revolutionary period [was] coming back" (Seifan 2011: 70).

Iraq, meanwhile, has moved from being a small-farmer, food-producing country to a major importer of food (FAO 2012). The capacity of agriculture to feed the population and play a role in the economy has been heavily affected by the poorly conceived, modernist policies of the Ba'ath regime, two Gulf wars, a protracted embargo, the US-led invasion, and years of insurgency. The UN oil-for-food program (1995–2010) undermined local production systems by flooding the market with cheap foodstuffs. The mainly Kurdish northern part of the country—de facto autonomous after 1991 and now referred to as the Kurdistan Autonomous Region (KAR)—was not directly affected by the US-led invasion and the following insurgency. However, the KAR authority neglected its agriculture and imported most of its food. When many investors withdrew after the rise of the Islamic State of Iraq al Sham (ISIS) in 2014, and relations between KAR and Baghdad soured, the vulnerability of this system of providing access for the population to basic foodstuffs became fully exposed.

Iraq became a food-importing country in the 1970s and is now externally dependent for 70 percent of its food supply. In 1990, after the imposition of sanctions, a public distribution system (PDS) for food was initiated. This provided households with food, avoiding a human crisis, and was further broadened under the oil-for-food program. While about 25 percent of the population is highly dependent on the public food-distribution system, another 35 percent is partially dependent on the PDS. In the KAR, the provision of food through cheap imports has been made public policy, disrupting local production.

For its part, Turkey has been engaged in a fundamental shift away from state-led models of economic development since the beginning of the 1980s. In the early 2000s, huge and wide-ranging cuts were made to the centralized support of its agricultural sector, with the scaling down and termination of state production, processing and distribution facilities, and marketing and market-protection mechanisms (Eşiyok 2004). Price liberalization resulted in lower returns and increased insecurity, and the privatization, diminishment, and liquidation of state-run institutions that had underpinned agriculture through subsidized inputs and crop-based support left farmers vulnerable in their relations with commercial trading organizations, retail chains, and industrial corporations. These neoliberal policies in agriculture were implemented by the government of Turkey but directed, facilitated, and enforced by international organizations such as the World Bank, IMF, and EU, resulting in an historically unparalleled squeeze on farming (Öztürk, Jongerden, and Hilton 2018).

Simultaneously, the downsizing of the state since the 1980s has restricted employment in the public sector, which has contributed to an excess supply of labor that has depressed real wages. This downsizing has sporadically acted as a causal factor in general economic slowdowns and crises, resulting in unemployment. Impairment of the labor unions further depressed wages and employment conditions. In 2006, just 13.3 percent of the workforce in Turkey was employed on the basis of collective labor agreements as compared with over 70 percent in most EU countries (Candaş et al. 2010: 70). Measured at 50 percent of the national median consumption expenditure, the figure for relative poverty is around 15 percent of the population (Yükseler and Türkan 2008: 105–106).

Clearly, the destruction of support measures and opening up of national food markets to international trade affected by the liberalizing thrust has contributed immensely to a disruption of agricultural production capacities and increased food insecurity in Iraq, Syria, and Turkey. Among the three, Turkey deserves particular attention since it remains one of biggest agricultural producers in the world, tenth by financial value, after China, India, the US, Indonesia, Brazil, Nigeria, Japan, Russia, and Argentina (İTB-Ar-GeM 2014: 3).

Declining agricultural production capacities in Turkey

Except in times of war, the Middle East and North Africa (MENA) region, as a whole, has been a relatively food secure zone. But recent years have witnessed structural shortcomings, negative tendencies, and weaknesses in the context of food security due to a combination of civil wars, unrests, and sanctions with neoliberal policies which, notwithstanding the variety of factors, point to a single narrative: a rise in poverty paralleling drastic changes in agricultural production, which has triggered food security issues throughout the entire region.

By looking at food production in terms of both availability–supply relations and accessibility–demand relations in Turkey, we see that Turkey is on the borderline with respect to food security; it is in between, one might say, as if in a kind of purgatory (*araf*); moreover, it has become one of the most striking examples of such in the region, perhaps the world. When we look at agricultural production in Turkey we see that, for quite a while now, input prices have been on the rise while product prices have been stable (so, declining in real terms), with climate changes (worsening aridity) causing erosion and leading to degradation of farmlands, all of which exacerbate already existing irrigation problems. This is why the case of Turkey puts out an urgent call.

The absolute poverty rate (food poverty)[6] in Turkey is around 0.5 percent, yet around 15 percent of the population is living in relative poverty and cannot access a sufficient quantity and quality of food, mostly because, since 1980, agricultural income has been declining in the country as a result of high input prices. The income-to-input relationship in farming can be indicated by the amount of wheat required to cover the cost of a liter of oil (diesel). As Figure 5.1 shows, this ratio increased steadily between 1998 and 2008, started to decline in 2009, and reached its peak in 2011, after which it became volatile—with the result that farmers in Turkey in 2017 were paying 3.9 times more for their oil (as compared to their income from wheat) than they were in 1998.

The price of fertilizers is the second (if not the first) most important regular input cost in agricultural production. Figure 5.2 shows that the prices of three most-commonly used fertilizers in Turkey were on the rise until 2015, except during the period of the economic crisis between 2008 and 2009. Figure 5.3 shows that producer prices of agricultural products also increased until

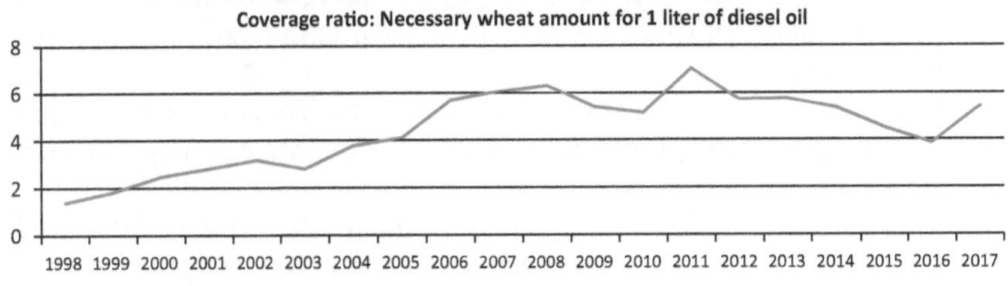

Figure 5.1 Diesel oil prices and wheat prices, 1998–2017.

Source: TURKSTAT, PPI Agricultural Sector, Tuik.gov.tr and for diesel oil, www.aytemiz.com.tr (visited 28.12.2017).

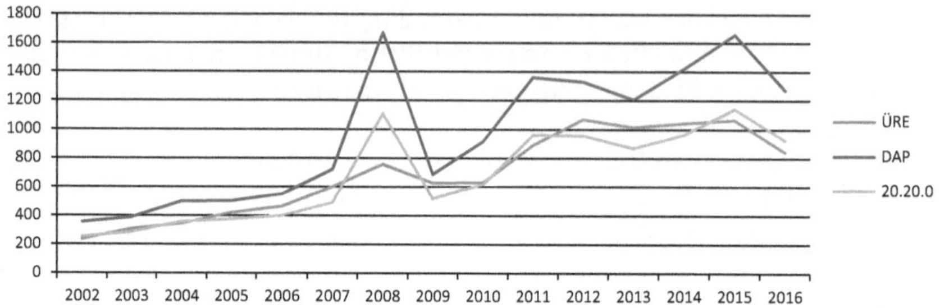

Figure 5.2 Prices of three fertilizers between 2002 and 2016.
Source: TURKSTAT. Tuik.gov.tr.

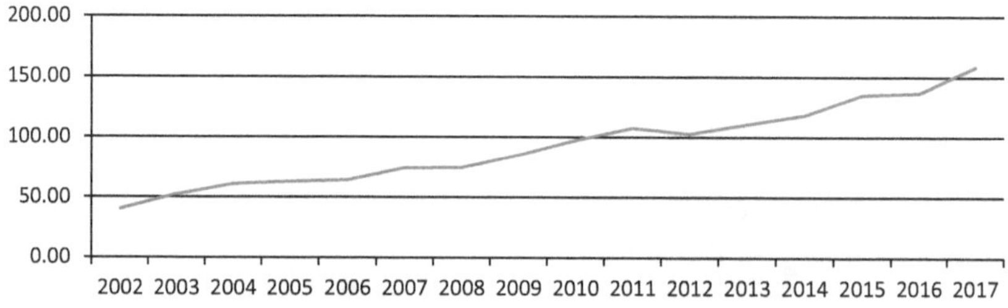

Figure 5.3 Producer price index of agricultural products between 2002 and 2017.
Source: TURKSTAT. Tuik.gov.tr.

2015, but subsequently, unlike fertilizer prices, they kept rising. Moreover, rises in both the producer price index and fertilizer price index rated close to each other while the price volatility of fertilizers was higher than that of agricultural products. The high price volatility of fertilizers as compared to agricultural product prices meant relatively uncertain costs, which resulted in financial issues for farmers, especially those without easy access to credit (essentially, smallholders with less than 10 ha of land, who own 78.9 percent of total farms and cultivate 34.3 percent of total land (Öztürk, Jongerden, and Hilton 2018).

Besides input prices, drought and irrigation inadequacy are also major problems for agricultural production in Turkey. Drought frequency and intensity have been rising over the last few decades. Except in the southeast, Turkey has relatively poor water sources; recent climate change associated with global warming, combined with ineffective water management practices, have worsened the situation for drought-stricken regions of the country. According to data provided by the State Water Office, Turkey has about 28 million ha of agriculture land; the irrigable land target is 8.5 million ha, but irrigated land only amounts to 5.9 million ha (DSİ 2017). Rises in irrigation capacity affect productivity and change the types of products cultivated. A general shift has occurred from traditional products (wheat, barley, etc.) to industrial production and crops used for animal feed (maize, clover, etc.) and to a specific preference in gardens for high value-added products (fruits, vegetables). Moreover, the construction of dams and ineffective irrigation practices came with salinization, reduced groundwater levels, and increased vulnerability to drought, not only in Turkey, but also in the wider region (Gleick 2014).

Table 5.1 shows total cultivatable agricultural land and types of usage in Turkey for the years from 1990 to 2016. The total amount of sown and fallow land has decreased, while land for vegetable and fruit growing has increased, resulting in a net total agricultural land decline of 15 percent. Arable land has been lost to the fast-developing construction sector, which includes the building of infrastructure (dams, roads) as well as houses and factories. Table 5.1, created by using the World Bank data set, clearly shows that urbanization has been favored in Turkey since 2007, at the cost of agricultural land.

Even though there has been a decline in total land usage, total agricultural production in the country has been rising in the last few decades, which shows that productivity has been increasing. However, this generalized increase in productivity has not been translated into sufficient yields for small-scale, family farms, as is evidenced by the number of farmers who could not afford to continue. Information gained from the Farmers' Record System[7] shows that the total number of active farm units in Turkey decreased from 2,588,666 in 2002 to 2,267,176 in 2016, representing a 12.4 percent decline over the last fourteen years (GTHB-BÜGEM 2017: 3). Some farmers have sought to ameliorate their financial difficulties by maximizing income through work as wage laborers and/or as entrepreneurs. Some of these farmers are employed in off-farm jobs while continuing with agricultural activities at home (for market

Table 5.1 Agricultural land of Turkey from 1990 to 2016, 1,000 ha

Agricultural land	*1990*		*2002*		*2016*		*Changes from 1990 to 2016*
Sown area	18,868	67.7	17,935	67.5	15,573	65.5	–17%
Fallow land	5,324	19.1	5,040	19	4,050	17	–24%
Area of vegetable gardens	635	2.3	930	3.5	804	3.4	27%
Area of fruits, beverage crops and spice	3,029	10.9	2,674	10.1	3,329	14	10%
TOTAL	**27,856**	**100**	**26,579**	**100**	**23,763**	**100**	**–15%**

Source: TURKSTAT. Tuik.gov.tr.

and/or self-sufficiency reasons), while others just give up and move to cities to survive (Öztürk et al. 2017).

In short, the number of farmers and the amount of cultivatable agricultural land in Turkey have been declining, despite an increase in productivity, because of high input prices, low and volatile agriculture product prices, poor climate, and various irrigation challenges. Thus we can conclude that environmental issues have negatively combined with neoliberal policies—specifically, diminishing agricultural supports, corporate monopolization of the global agricultural input and food sectors, and the effects of these corporations on agricultural-based food supply, sales, and farmer and consumer practices, all of which are related to food security and food sovereignty (Patel 2009; Agarwal 2014). Unsurprisingly, therefore, official statistics indicate a significant net migration from rural to urban areas (Öztürk 2012b: 141).

Three important caveats in respect to this urbanizing shift should be noted, however. First, while rural-to-urban migration has led to a decline of the rural population, a movement in the opposite direction occurred during the periods from 1975 to 1980 and from 1995 to 2000, when there were net increases in the rural population because of the rate of natural increase. Second, this opposite movement did not occur in the southeast of Turkey, which was heavily affected by a forced evacuation of villages as part of a counter-insurgency program to combat the Kurdistan Workers Party (PKK). Third, there was a decrease of rural-to-rural migration, and an increase of urban-to-urban migration, the latter accounting for about half of the migration in the period covered (Öztürk 2012b: 140–141).

Turkey remains a major world agricultural producer in the sense that it produces a large quantity and wide range of items for both national and

Table 5.2 The poverty rates according to poverty line methods, 2002–2014 in Turkey

Methods	*Percentage of poor individuals (%)*				
	2002	*2007*(3)	*2008*	*2009*	*2014*
Food poverty (4)	1.35	0.48	0.54	0.48	
Complete poverty (food+nonfood) (4)	26.96	17.79	17.11	18.08	
Below $ 2.15 per capita per day (1)	3.04	0.52	0.47	0.22	0.03
Below $ 4.3 per capita per day (1)	30.30	8.41	6.83	4.35	1.62
Relative poverty based on expenditure(2)(4)	14.74	14.70	15.06	15.12	

Source: TurkStat, Poverty Study, 2014.

Notes:

(1) Here the current PPP values in TL are used for the equivalents of $ 1.
(2) It's based on the 50% of equalized median consumption expenditure.
(3) New population projections are used since 2007.
(4) Values are not calculated due to the methodological revision studies since 2010

international markets with increasing productivity rates. Its agricultural sector has been structurally adjusted to the world economy, having become an active participant in the world economy in terms of importing and exporting agricultural products (Aydın 2010; Öztürk 2012b; Keyder and Yenal 2013; Pamuk 2014). Nevertheless—and as a function of this development—government policies, land degradation and non-agricultural usage of lands, worsening climate conditions, and the negative impact of all these in particular on small-scale farming,[8] have combined to result in a diminishing production capacity among small-scale farmers and thus increasing food insecurity.

Another problem in Turkey is the high transaction costs of agricultural and food products. The gap between producer and consumer prices is high, especially when it comes to vegetables and fruits, due to transportation and storage costs and also to food waste. Regarding the latter, we can say that as Turkey became an active participant of the global economy over the last 30 years, so did it also become an active participant in the global system of food-waste regimes (Cloke 2013; Gille 2012).[9] The loss of food mainly occurs during transportation, storage, and sale. The uncertainty pre-purchase loss creates is a crucial problem for Turkey's producers. As such, food loss put additional burdens on small-scale producers, the dominant form of agricultural production in Turkey. Apart from the cost of food waste and loss, transportation, and storage, other major reasons for high consumer prices in Turkey include foreign demand, commissions, and speculation, which all bear directly on food accessibility and poverty.

Food accessibility and the poverty rate

According to official statistics, about three million (14%) of Turkey's households received social aid and 656,000 (3%) received food aid in 2016.[10] The (complete) poverty rate (food+nonfood) was 28.12 percent in 2003 and 17.11 percent in 2008, but 18.08 percent in 2009 (see Table 5.2). Unfortunately, TurkStat, the state body responsible for official statistics, has not presented complete figures on poverty since 2009. According to income and food-purchasing indicators from before and since 2009, however, it can be ascertained that the 2009 and present figures for poverty are not much changed. Thus, poverty reduction in Turkey appears to have stalled over the last decade.

The poverty rate of a country indicates the proportion of poor people who, for various reasons, have difficulty in accessing sufficient food to live healthily and to biologically reproduce. Thus, Turkey's contemporary lack of food self-sufficiency has started to pose problems, which is evident particularly when we compare its poverty to absolute poverty rates: less than 1 percent and about 18 percent, respectively. The former indicates the proportion of citizens in Turkey who cannot access food, while the latter shows those who may be having problems in acquiring enough food of sufficient quality. Even though absolute poverty totals less than 1 percent, when measured at 50 percent of the national median consumption expenditure, the figure for relative poverty is around 15 percent. This is even higher in villages, where the disparity between urban and rural areas means that relative poverty stands at about 35 percent of the rural population according to poverty statistics of TurkStat (Öztürk 2012b: 184–186).

There can be little doubt that high food prices and the stalled poverty-rate reduction indicate significant problems with respect to food production and consumption in Turkey, particularly in rural areas, and that this should be taken into consideration by critical food studies. That this has not featured in the literature and research to date constitutes a major omission. Perhaps the oversight is due to Turkey's renown for self-sufficiency in agriculture, colored with the analytical difficulties involved. Here we argue that, while high prices negatively affect food accessibility much more than production, production itself is fragile because of high prices of inputs and the cost of waste, transportation, and storage, particularly in a country like Turkey, where small-scale producers dominate agrarian structures. This, in the end, inevitably transforms food accessibility into a structural problem.

Food accessibility and adaptation behavior

We can see the weakness of food accessibility in Turkey by combining the figures for poverty, income inequality, and the food-purchasing tendency of households. Tables 5.2, 5.3, and 5.4 provide information about relative poverty as well as about household poverty, which denotes relative poorness among households in a given category.[11] According to the Income and Living Conditions Survey 2018 of TurkStat, if someone who has 40 percent of equalized household income is accepted as poor, the number of poor people is almost 6.5 million, some 8.4 percent of the total national population in 2016. If the poverty threshold rate is set at 50 percent of equalized household income, then the poverty rate reaches 14.3 percent, while for 60 percent it is 21.2 percent. Overall, an overview of poverty-threshold numbers suggests that the proportion of households with nutritional needs constitutes a structural problem.[12]

Taking a closer look at poverty, we can analyze the distribution of income. We already know that Turkey has one of the most unequal income distribution figures in the world. Table 5.4 shows that the bottom 20 percent by income receive only about 6 percent of the total GDP, whereas the top 20 percent gains 46 percent. Obviously, the lowest income group's access to various commodities, including food, is lower than any other group's, making this group dependent on state social policies.

Figure 5.4 shows that people in the highest income group in Turkey spend about three times more on food and beverages than do those in the lowest income group, who, it can be deduced, would purchase around three times more than the current level if they had the necessary income. According to the Engel Rule: When income increases, food-purchasing decreases as a

Table 5.3 Number of the poor, poverty rate and poverty gap by equivalized household disposable income, 2006–2016

	Years		
	2006	*2011*	*2016*
%40			
Poverty threshold (TL)	1,881	3,255	5,693
Number of poor (Thousand person)	8,665	7,288	6,482
Poverty rate (%)	12.8	10.1	8.4
Poverty gap[(1)]	29.1	25.4	20.8
%50			
Poverty threshold (TL)	2,351	4,069	7,116
Number of poor (Thousand person)	12,548	11,670	11,026
Poverty rate (%)	18.6	16.1	14.3
Poverty gap[(1)]	31.7	26.3	24.3
%70			
Poverty threshold (TL)	3,291	5,697	9,963
Number of poor (Thousand person)	21,611	21,730	22,096
Poverty rate (%)	32.0	30.0	28.7
Poverty gap[(1)]	34.8	31.0	28.5

Source: TURKSTAT, Income and Living Conditions Survey. Reference period of incomes is the previous calendar year.

Note:

(1) Poverty gap ratio informs about poverty level. It represents the severity of poverty and it is too much if it approaches "100" and it represents the poverty risk degree and it is low if it reduces. Poverty gap formula is:

Poverty gap = ((Poverty threshold-Median income of poor by EII)/Poverty threshold)*100

Table 5.4 Distribution of annual household disposable income by quintiles ordered by household disposable income, 2006–2016

Years	*Quintiles*				
	First quintile	*Second quintile*	*Third quintile*	*Fourth quintile*	*Last quintile*
2006	5.8	10.5	15.2	22.1	46.5
2011	6.5	11.0	15.5	21.9	45.2
2016	6.3	10.6	15.2	21.6	46.3

Source: TURKSTAT, Income and Living Conditions Survey.

Note:

Figures in table may not add up to totals due to rounding. Reference period of incomes is the previous calendar year. Households are sorted in ascending order by household disposable income and divided into 5 equal parts, the bottom income group is defined as "the first quintile" and the top income group as "the last quintile".

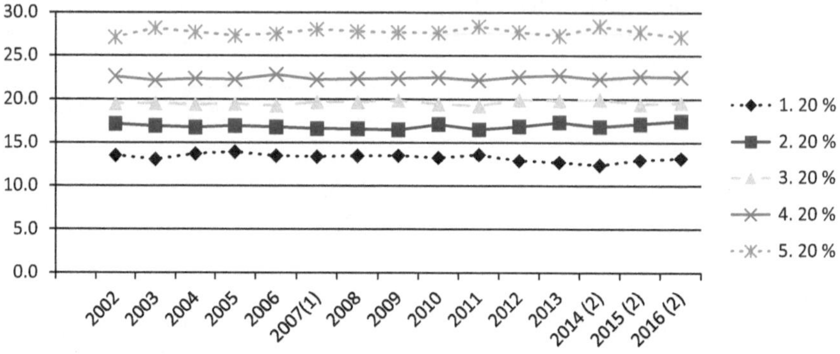

Figure 5.4 Shares of quintiles ordered by income in food and non-alcoholic beverages expenditures, Turkey, 2002–2016.

Source: TurkStat, Household Budget Survey.

Notes:

(1) Note that new population projections have been used since 2007.

(2) It is based on the new administrative division for sampling design since 2014.

proportion of total income, even if absolute food purchasing rises. Figure 5.4 shows that low-income group food purchasing is relatively *less* than that of higher income groups. Therefore, there is clearly a huge food-purchasing gap between low- and high-income groups. This provides further evidence for the inability of lower income groups to afford enough good food.

Considering income level and accessibility for the lowest income group (poorest segment) in absolute figures, we see that the mean household annual disposable income for Turkey in 2016 was 41,399 TL (10,981 USD)[13] and the lowest household group disposable income mean 12,957 TL (3,437 USD). According to Turk-İş (the country's largest labor union), the hunger threshold in 2016 was 1,432 TL (380 USD) monthly for a family of four, the average household size in Turkey.[14] This represents 17,184 TL annually, but the mean household income of the lowest group for disposable income is 12,957 TL per year. Again, these numbers demonstrate the considerable weakness of the lowest income group's accessibility to food. Table 5.5 provides further evidence for this: 62.9 percent of the lowest income households, defined as those below 60 percent of average income, cannot afford a meal with meat, chicken, or fish every second day.

Table 5.6 presents food purchasing in a socio-economic status (SES) format, where SES group definition is based on economic and social variables such as education level and occupation. Let us consider the SES group simply as indicating income level, which decreases from the AB to DE groups. The poorest (DE) group food-consumption level in 2016 was lower than that in 2006, even as the average consumption level rose 10 percent in the same period. Another important indication from these figures concerns the food-purchasing fragility of the poor. During the economic crisis period from 2008 to 2009, total

Table 5.5 Distribution of non-institutional population by equalized household disposable median income groups and specified living condition indicators, 2006–2016

Non-institutional population and living conditions indicators – Capacity to afford a meal with meat, chicken or fish every second day	*Total*	*Income group*		
		Below 60% of the median income	*Between 60%–120% of the median income*	*Above 120% of the median income*
2006				
Can afford (%)	**35.9**	10.5	25.3	60.9
Cannot afford (%)	**64.1**	89.5	74.7	39.1
2011				
Can afford (%)	**39.8**	13.6	30.6	63.9
Cannot afford (%)	**60.2**	86.4	69.4	36.1
2016				
Can afford (%)	**62.3**	37.1	56.1	82.3
Cannot afford (%)	**37.7**	62.9	43.9	17.7

Source: TURKSTAT, Income and Living Conditions Survey.

Note:
Figures in table may not add up to totals due to rounding.
Note that the reference period for incomes is the previous calendar year, denoting magnitude null. In accordance with Eurostat's recommendations, here some revisions were made so that the questions directed be easily comprehended and accurately understood by the respondents. (http://tuik.gov.tr/PreTablo.do?alt_id=1011, visited, 10.01.2018)

Table 5.6 Food consumption index by SES group 2006–2016

Years	*AB*	*C1*	*C2*	*DE*	*Food*
2006	100%	100%	100%	100%	100%
2007	100%	102%	101%	93%	98%
2008	102%	102%	101%	88%	96%
2009	108%	111%	110%	90%	102%
2010	110%	117%	113%	94%	106%
2011	113%	119%	115%	93%	107%
2012	114%	117%	112%	95%	107%
2013	118%	121%	114%	92%	108%
2014	116%	120%	116%	93%	108%
2015	116%	122%	122%	100%	113%
2016	112%	119%	120%	98%	110%

Source: Ipsos-Turkey, Data derived by Ipsos-Turkey household consumer panel experts for this study. Researchers are grateful to Ipsos and its experts. For their invaluable help.

Information for Ipsos: www.ipsos.com/en

Table 5.7 Homemade food supply by SES (socio-economic-status) groups, volume share

Years	*AB*	*C1*	*C2*	*DE*	*Total*
2016	6.52%	15.91%	22.53%	55.05%	100.0%
2017	7.5%	16.1%	22.5%	53.9%	100.0%

Source: Ipsos-Turkey. Data derived from Ipsos household consumer pan Data derived by Ipsos-Turkey household consumer panel experts for this study.

Table 5.8 Rural household food supply 2012 and 2016, %

	N:436 Household, 2012 (1)				*N:1222 Household, 2016 (2)*			
	Self-home produced (I)	*Bought*	*Both (III)*	*I+III*	*Self-home produced (I)*	*Bought*	*Both (III)*	*I+III*
Bread	53	26	21	74	34.9	40.7	24.5	59.4
Cheese /Butter	49	45	6	55	30.3	65.1	4.7	35
Eggs	51	42	7	58	38.5	56	5.5	44
Milk	55	41	4	59	36.8	61.5	1.7	38.5
Pickle	71	25	4	75	58.4	38.1	3.5	61.9
Sauce	52	44	5	57	47.5	48.9	3.5	51
Tarhana (Soup Base)	52	46	2	54	42.6	54.7	2.8	45.4
Cracked wheat	48	50	2	50	30.8	67	2.2	33
Meat*	13	82	5	18	10.4	85.8	3.8	14.2
Vegetables and Fruits	28	60	12	40	31	55.1	13.9	44.9

Source: (1) Öztürk, el all. 2014, (2) Kentel, F. 2017.

Note:

* Meat is traditionally regarded as a luxury food in most areas in Turkey and accordingly rarely consumed.

food purchasing in Turkey decreased from the benchmark of 100 to 96; this was because DE group food purchasing decreased from 100 to 88 (while the other SES groups' food purchasing did not decrease).

Another important factor to consider is the supply of food grown by the household. People in Turkey prefer food that they grow themselves, and for various reasons, including its taste (traditional), quality (natural, healthy) or cost (cheap), with the latter being particularly attractive for low-income households. As Table 5.7 shows, low-income households consume more than half of the food grown by households in Turkey.

Table 5.8 shows the contemporary situation regarding the homemade food supply in villages, which is obviously more extensive in rural than urban areas due to relative ease of access (farming households also grow food that they eat). Data compiled from two research projects conducted in 2012 and 2017 show that, excluding vegetables and fruits, the amounts of all types of home-prepared/produced food have been decreasing. This decrease appears to have been ongoing since the 1980s (Öztürk et al. 2014; Kentel 2017). One important characteristic of homemade food is that people share it with relatives, the poor, and the disabled. It is one of the most powerful instruments in social life, providing connection and solidarity among rural communities. Given that homemade food is generally cheaper than manufactured and packed food, the decline in homemade food might also represent a decrease in poor people's access to food.

Conclusion

While failed modernization policies, war, and cheap food imports have made Iraq and Syria food insecure, there is a bidirectional tendency in food security issues in Turkey. Even though this country has a strong food-production capacity, there are some risks and structural limits for food production and accessibility.

Currently, Turkey produces and exports more food than it used to, but it has been losing agricultural lands to erosion, aggressive development of construction, and the flight of rural populations to the cities. Neoliberal economic policies have weakened the position of small- and medium-size traditional farms across Turkey by forcing the agricultural sector to produce for the competitive world market and sell at uncertain prices. This not only means lower product prices over the years but also overall economic insecurity as a result of price fluctuation.

All of these factors exacerbate the risks and structural problems on the food supply side in Turkey. Unequal income distribution and high poverty rates are indicative of other structural problems, such as urban-rural differentiation and intraregional differences that the country has been encountering in food accessibility during recent years.

In the wider region, in particular Syria, we have seen how a push toward irrigated agriculture in combination with subsequent years of drought has negatively affected production capacity as well as the opening of markets; thus the slashing of subsidies for farmers has contributed to increased poverty and insecurity among the rural population. As cities grow faster than ever in the region, so do the numbers of urban poor resulting from the massive rural-to-urban flight over the last 30 years. In this context, it is important to stress that the social-aid programs initiated in the last decades have helped the urban poor, particularly in Turkey, but they have never been sufficient.

As Weis (2010) puts it, the solution cannot just be industrial farming, which poses serious environmental problems while promising cheaper food. Given that putting burdens on small-scale farming triggers rural-to-urban flight and increases unemployment rates and thus poverty, giving up on neoliberal agricultural policies and supporting small-scale farming cannot be considered just an ethical position; on the contrary, it is the solution to the food security issues in Turkey and in the region.

Notes

1 The authors are grateful to the editors of the present volume for their invaluable insights and assistance. They would like to extend their special thanks to Andy Hilton and Oguz Can Ok for their continued support. They are indebted to IPSOS and its experts, whose assistance was of immeasurable help.

2 "Iraq's economy was mainly based on agriculture before the development of the oil sector in the 1930s. As oil revenues rose, the country experienced an urbanization process that led to higher population growth and a stagnation of the agricultural sector. In the 1960s, the demand for food began to exceed supply, and Iraq was compelled to increase its reliance on food imports. Iraq's involvement in armed conflicts, beginning with the Iran–Iraq war in 1980–1988, severely affected the country's food security" (Eklund 2017:2); despite urbanization, the majority of the poor still live in the countryside and are more likely to be identified through their sector of employment, with poor household heads tending to work in agriculture and fishing and in the construction, transport, storage, and communication sectors.

3 For more information on the term "food security," including the term "food sovereignity," see Bina Agarwal (2014) and March Edelman, Tony Weis, Amita Baviskar et al. (2014). Also see www.fao.org/docrep/005/y4671e/y4671e05.htm#TopOfPage (retrieved August 15, 2018).

4 The critical literature on food studies is univocal in regarding the problem with food security as less about the quantity of food produced and more about its quality and accessibility, the latter point having been emphasized as, in fact, the main problem following the research in India by Amartya Sen (1981).

5 IPSOS is an international market research company. For more information see www.ipsos.com/en (retrieved August 15, 2018).

6 Here the current PPP value in TL used for the equivalent of 1 USD per day.

7 This a national database maintained by the Ministry of Agriculture and Forestry of the Republic of Turkey.

8 When small-scale farmers lose income because of market and/or natural conditions, they tend to seek alternative income sources while continuing with farming (Boratav 2004: 110). This is a typical strategy for small-scale farms in Turkey when economic conditions are against agriculture in general and family farms in particular. That is to say, small-scale farm families have more capacity than it might seem. This would also include small-scale family farms in the mountains, where agriculture is limited by nature. For a recent case study on family farms in a mountainous area, located in western Turkey, and their resilience and capacity to cope with neoliberal economic policies in terms of agrarian transformation (see Gür 2018).

9 Zsuzsa Gille in her informative article (Gille 2012: 27) utilizes the concept of waste regime in order to understand "the global connections involved in generating food waste. This concept treats waste as a social relationship and assumes that in any economy there is a waste circulation in addition to a value circulation, and that the two are interdependent. For this reason, the author critiques metaphors, such as value chains or supply chains, that have dominated the scholarship on food and agriculture."

10 For total household number, see TurkStat, "Statistics on Family" (2016); for social aid, see www.takvim.com.tr/ekonomi/2017/03/09/3-milyon-haneye-yardim-bakanlik-acikladi (retrieved August 15, 2018).
11 These households have income less than medium income if the threshold is 40 percent.
12 For more information on "Income and Living Conditions Survey" of TurkStat (2018), see www.tuik.gov.tr/PreTabloArama.do?metod=search (retrieved August 15, 2018).
13 1$=3.77 TL (the currency exchange system in Turkey has been free-floating since 1980).
14 www.turkis.org.tr/ARALIK-2016-ACLIK-ve-YOKSULLUK-SINIRI-d1338 (retrieved August 15, 2018).

References

Abboud, Samer N. 2013. "Economic transformation and diffusion of authoritarian power in Syria." In *Democratic Transition in the Middle East: Unmaking Power*. Ed. Larbi Sadiki, Heiko Wimmen, and Layla Al-Zubaidi. 159–177. London: Routledge.

Agarwal, Bina. 2014. "Food security, food sovereignty and democratic choice: Addressing potential contradictions." *Journal of Peasant Studies* 41(6): 773–784.

Al-Hindi, Atieh. 2011. "Syria's agricultural sector: Situation, role, challenges and prospects." In *Agriculture and Reform in Syria*. Ed. Raymond Hinnebusch, Atieh al-Hindi, Munzer Khaddam, and Myriam Ababsa. Fife, Scotland: University of St Andrews Centre for Syrian Studies.

Aydın, Zülküf, 2010. "Neo-liberal transformation of Turkish agriculture." *Journal of Agrarian Change* 10(2): 149–187.

Barnes, Jessica. 2009. "Managing the waters of Ba'ath country: The politics of water scarcity in Syria." *Geopolitics* 14(3): 510–530.

Beer, Sam. 2016. *The United States' Program for Agriculture in Post-Invasion Iraq*. New York: Angelica.

Bernstein, Henry, Ben Crow, Maureen Mackintosh, and Charlotte Martin. 1990. *The Food Question: Profits Versus People?* London: Earthscan.

Boratav, Korkut. 2004. *İstanbul ve Anadolu'dan Sınıf Profilleri* [Class Profiles from Istanbul and Anatolia]. 2nd ed. Ankara: İmge.

Candaş, Ayşen, Yılmaz Volkan, Günseli Sevda, and Y. Ç. Burcu. 2010. Türkiye'de Eşitsizlikler: Kalıcı Eşitsizliklere Genel Bir Bakış. İstanbul: Boğaziçi Üniversitesi Sosyal Politika Forumu.

Chatel, Francesca de. 2014. "The role of drought and climate change in the Syrian uprising: Untangling the triggers of the revolution." *Middle Eastern Studies* 50(4): 521–535.

Cloke, Jon. 2013. "Empires of waste and the food security meme." *Geography Compass* 7(9): 622–636.

DSİ (Devlet Su İşleri). 2017. *Toprak Su Kaynakları* [Land Water Resources]. Accessed January 09, 2018, www.dsi.gov.tr/toprak-ve-su-kaynaklari

Edelman, Marc, Tony Weis, Amita Baviskar, Saturnino M. Borras Jr, Eric Holt-Giménez, Deniz Kandiyoti, and Wendy Wolford. 2014. "Introduction: Critical perspectives on food sovereignty." *Journal of Peasant Studies* 41(6): 911–931.

Eklund, Lina, Abdulhakim Abdi, and Mine Islar. 2017. "From producers to consumers: The challenges and Opportunities of agricultural development in Iraqi Kurdistan." *Land* 6(44): 1–14.

Eşiyok, B. Ali. 2004. *Kalkınma Sürecinde Tarım Sektörü: Gelişmeler, Sorunlar, Tespitler ve Tarımsal Politka Önerileri.* Ankara: Türkiye Kalkinma Bankasi A. Ş., Arastırma Müdürlügü.

FAO (Food and Agriculture Organization). 2001. Sarris, Alexander. 2001. *Agricultural Development Strategy for Syria.* Damascus: FAO.

FAO (Food and Agriculture Organization). 2012. *Iraq: Agriculture Sector Note.* Rome: FAO.

Friedmann, Harriet, and P. McMichael. 1987. "Agriculture and the state system: The rise and fall of national agricultures, 1870 to the present." *Sociologia Ruralis* 29(2): 93–117.

Gibson, Glen R. 2012. *War and Agriculture: Three Decades of Agricultural Land Use and Land Cover Change in Iraq*, Unpublished PhD dissertation in Geospatial and Environmental Analysis (Virginia Polytechnic Institute and State University: Blacksburg)

GTHB-BÜGEM (Gıda Tarım ve Hayvancılık Bakanlığı Bitkisel Üretim Genel Müdürlüğü). 2017. *Bügem Faaliyetleri* [Activities of Bügem]. Ankara.

Gille, Zsuzsa. 2012. "From risk to waste: Global food waste regimes." *Sociological Review* 60(52): 27–46.

Gleick, Peter H. 2014. "Water, drought, climate change, and conflict in Syria." *Water, Climate and Society*, 6(3): 331–340.

Gobat, Jean, and Kristina Kostial. 2016. *Syria's Conflict Economy*, IMF Working Paper 16/123.

Gür, Faik. 2018. "Yunt Dağı Köylerinde Kırsal Dönüşüm" [Agrarian transformation in the villages of Yunt Mountain]. In *Sürüdürülebilir Yaşam Penceresinden Yerel ve Kırsal Kalkınma.* Ed. Faik Gür and Bayramoglu Murat. Istanbul: Ozyegin University Press.

IFAD (International Fund for Agricultural Development). 2001. *The Syrian Arab Republic Country Programme Evaluation Evaluation Report,* www.ifad.org/documents/38714182/39715119/syria.pdf/33d9865a-5344-42e0-ad80-eaa7ff0592ee

IME (Institution of Mechanical Engineers). 2013. *Report: Global Food: Waste Not Want Not.* www.imeche.org/policy-and-press/reports/detail/global-food-waste-not-want-not.

İTB-Ar-GeM (İzmir Ticaret Borsası Ar-Ge Müdürlüğü). 2014. *Rakamlarla Dünya Tarımı.* Izmir.

Joya, Angela.2010. "A Comparative Study of Neoliberalism in Egypt and Syria." In *Confronting Global Neoliberalism: Third World Resistance and Development Strategies.* Ed. Richard Westra. Atlanta: Clarity Press.

Kentel, Ferhat. 2017. *Kır Mekânının Sosyo-Ekonomik ve Kültürel Dönüşümü: Modernleşen ve Kaybolan Geleneksel Mekânlar ve Anlamlar*. Unpublished Research Project report.

Keyder, Çağlar, and Zafer Yenal. 2013. *Bildiğimiz Tarımın Sonu: Küresel İktidar ve Köylülük* [The End of Agriculture We Know: Global Power and Peasantry]. Istanbul: İletişim.

Koc, Mustafa, Jennifer Sumner, and Anthony Winson. 2017. *Critical Perspectives in Food Studies.* 2nd ed. Oxford: Oxford University Press.

Lund, Aron. 2014. *Drought, Corruption, and War: Syria's Agricultural Crisis, Carnegie Middle East Center*. Retrieved August 11, 2018, http://carnegie-mec.org/diwan/55376.

Moore, Pete, and Christopher Parker. 2007. "The war economy of Iraq." *Middle East Report* 37(3): 6–15.

Öztürk, Murat. 2012(a). *1980 Sonrası Yıllarda Türkiye'de Tarımda ve Kırda Dönüşüm Dinamikleri*. Unpublished Research Report.

Öztürk, Murat. 2012(b). *Agriculture, Peasantry and Poverty in Turkey in the Neo-Liberal Age*. Wageningen: Wageningen Academic Publishers.

Öztürk, Murat, Joost Jongerden, and Andy Hilton. 2014. "Commodification and the social commons: Smallholder autonomy and rural-urban kinship communalism in Turkey." *Agrarian South: Journal of Political Economy* 3(3): 337–367.

Öztürk, Murat, Joost Jongerden, and Andy Hilton, 2018. "The (re)production of the new peasantry in Turkey." *Journal of Rural Studies* 61: 1–11. doi:10.1016/j.jrurstud.2017.10.009.

Öztürk, Murat, Besir Topaloglu, Andy Hilton, and Joost Jongerden. 2017. "Rural-urban mobilities in Turkey: Socio-spatial perspectives on migration and return movements." *Journal of Balkan and Near Eastern Studies* 20(5): 513–530. doi:10.1080/19448953.2018.1406696.

Pamuk, Şevket. 2014. *Türkiye'nin 200 Yıllık İktisadi Tarihi*. İstanbul: Iş Banakası Kültür Yayınları.

Patel, Raj. 2009. "Grassroot Voices." *Journal of Peasant Studies* 36(3): 663–706.

Rocchi, Benedetto, Donato Romano, and Raid Hamza. 2013. "Agriculture reform and food crisis in Syria: Impacts on poverty and inequality." *Food Policy* 43(C): 190–203.

Schnepf, Randy. 2004. *Iraq Agriculture and Food Supply: Background and Issues*. CRS Report for Congress. Washington, DC: Congressional Research Service.

Seifan, Samir. 2011. *The Road to Economic Reform in Syria, St Andrews Papers on Contemporary Syria*. Fife: University of St Andrews Centre for Syrian Studies School of International Relations.

Sen, Amartya. 1981. *Poverty and Famines: An Essay on Entitlement and Deprivation*. New York: Oxford University Press.

Senses, Fikret. 2016. Turkey's experience with neoliberal policies since 1980 in retrospect and prospect. *New Perspectives on Turkey* 47: 11–31.

Weis, Tony. 2010. "The accelerating biophysical contradictions of industrial capitalist agriculture." *Journal of Agrarian Change* 10(3): 315–341.

Yükseler, Zafer, and Ercan Türkan. 2008. *Türkiye'de Hane halkı: İşgücü, Gelir, Harcama Ve Yoksulluk Açısından Analizi*. [Household in Turkey: Labor, Income, Expenditure and Poverty Analysis]. İstanbul: TÜSİAD.

6 Links between land access, land use, and hunger in today's neoliberal Nicaragua

Birgit Schmook, Lindsey Carte, Claudia Radel, and Santana Navarro Olmedo

Introduction

In this chapter, using as an example a selection of rural communities in the municipality of Somoto, located in the northwest Nicaraguan department of Madriz, we demonstrate how land access and land tenure are intricately linked to smallholder food insecurity. In Nicaragua, the great heterogeneity of land-access arrangements and land-tenure systems has complicated food production, especially for smallholders and the landless, groups that often have access to very little land with which to produce food for their families. Today there is a re-concentration of land in the hands of large landholders, while access to land for small-scale production continues to include the tenure of small parcels and an array of rental, borrowing, and share-cropping arrangements. The complexity of the land system is rooted in both colonial and recent history, during which Nicaragua suffered 43 years of dictatorship under the Somozas; was governed by the leftist Sandinista revolutionary government during the 1980s; transitioned to centrist-neoliberal governance between 1990 and 2006; and has been ushered back in the present to an authoritarian populist leftist regime under Daniel Ortega (Martí i Puig and Baumeister 2017).

Throughout this history, land access for peasants has been a persistent problem. Despite undeniable improvements regarding land access and distribution during the Sandinista years, the present-day situation is again characterized by highly unequal land distribution and land concentration. Despite a number of poverty-alleviation programs implemented by the current government, especially among the rural population, elevated poverty levels have combined with a neoliberal macroeconomic policy that still favors agroexports and limits access to credits and agricultural inputs for small-scale farmers, thus placing additional hardships on smallholders and the landless. The resulting insufficient food-crop production puts land-poor farmers and their families at risk of seasonal periods of hunger, with decreases in overall dietary diversity and caloric intake.

Recent research highlights the importance of the relationship between land tenure and food security. In particular, it illustrates three areas: first, the limits of land reform for producing food security (Saturnino, Franco, and Suárez 2015); second, the problems inherent to creating effective food sovereignty and security policy under neoliberalism (Trauger 2014; Godek 2015); and, third, the challenges to procuring food security even from within fair-trade structures with the stated goal of bolstering sustainable livelihoods (Bacon et al. 2014; Bacon 2015). Our Nicaraguan case study contributes to this literature through our focus on the linked land-tenure and food security experiences of two of six study villages, which illustrate the complexity of contemporary land-access arrangements in Nicaragua. Each of these two communities represents a different trajectory-type and set of small-scale farmer responses that are widespread across Nicaragua: a former Sandinista cooperative and a remote village caught in feudal agrarian structures, both suffering from severe lack of food. We discuss how land-tenure and access regimes have contributed to food insecurity in these villages and demonstrate that, although the details of local trajectories can vary from village to village, the experience of food insecurity has a common cause—inequities in land assets and a lack of entitlements in a neoliberal context. This chapter highlights that, regardless of political alignments or current land tenure arrangements, food insecurity emerges as a transversal problem among smallholders and the rural landless.

Background: national conditions affecting small-scale production

In this section we focus on two aspects of government policies and programs across Nicaragua that have made an impact on small-scale production in the Somoto region. The first is the relationship between food security/insecurity in agricultural production, and the second is land tenure.

Food security/insecurity and agricultural production

In 1996, the World Food Summit defined food security as a state in which "all people at all times have access to sufficient, safe, nutritious food to maintain a healthy and active life" (FAO 1996; UN 2012). Most of these conditions do not apply in Nicaragua, which is the second poorest country in Latin America after Haiti (IFAD 2012), has a persistent food deficit (WFP 2018), and ranked 21 out of 79 on the 2012 Global Hunger Index (IFPRI 2012). Poverty is largely a rural problem in Nicaragua, with about half of the population living in rural areas and 25 percent of farming households classified as extremely poor and experiencing chronic or temporary food insecurity (IFAD 2012; FAO 2013). Among the most food insecure areas is the department of Madriz, where Somoto is located (WFP 2018).

In 1980, 55 percent of the population suffered from undernourishment, but Nicaragua since has made the most headway in Central America. By 2010, Nicaragua had achieved the Millennium Development Goal of halving the proportion of people suffering from hunger (FAO 2015). The increasing production of staple food crops helped to achieve this goal, but a significant amount of Nicaragua's available food continues to come from food aid, and per capita food assistance is comparatively high (Sahley et al. 2005; WFP 2018).

According to a recent USAID-funded food security assessment, the three principal reasons for significant lack of food access in Nicaragua are (1) a highly skewed distribution of income, (2) a highly skewed distribution of land, affecting the poor in the rural areas, and (3) a low level of national and per capita income (Sahley et al. 2005). Families relying on subsistence farming, on very limited land, have a difficult time getting out of poverty and producing enough food. Insufficient investment in general infrastructure, such as roads and electricity, as well as in agricultural infrastructure such as irrigation and storage facilities, also limits opportunities (Sahley et al. 2005). The state's continued orientation toward export crops, with disinvestments in the small-scale food-production sector, has further diminished the capacity for meeting basic food requirements (HabibMintz 2004; Martí i Puig and Baumeister 2017).

Agriculture, forestry, and fisheries together contribute about a third of Nicaragua's annual GDP and roughly half its total export revenue (BCN 2015; World Bank 2015). The agricultural economy in Nicaragua can be divided into two main sectors, with some overlap between them: a staple-crop sector and a cash-crop sector. The staple-crop sector, which produces rice, maize, and beans, comprises mostly small, family-run farms. Most staple crop farmers have minimal access to land, to inputs, and to improved seeds, and their farms are often located in areas with poor soils and difficult access to markets. While yields in Nicaragua's cash-crop sector have kept pace with world averages, yields for maize, a staple crop, have grown only at half the global average rate from 1961 to the present, and bean yields have actually fallen slightly over this same period. Increases in overall maize and bean production of roughly 300 percent, in the context of a roughly 200 percent growth in population since 1960, have mainly been achieved by expanding the area of land under cultivation (Gourdji et al. 2015).

The transition to the Chamorro government in 1990 brought neoliberal reforms that reduced credit and extension services for smallholders, making food production increasingly difficult. These policies have been continued under Ortega's government, in place since 2007; the new rural bourgeoisie, established during the Chamorro government, has become the agricultural/landed elite linked to national and international agroindustry (Martí i Puig and Baumeister 2017).

Land tenure

Land concentration was pervasive before the Sandinista revolution in 1979 and only briefly interrupted by the Sandinista years of the 1980s, with the implementation of a structural agrarian reform and significant investment in the agricultural sector (Martí i Puig and Baumeister 2017). Nevertheless, the Sandinista government's agricultural policies largely prioritized production modernization, often at the expense of smallholders and landless peasants, who did not receive the necessary financial and technical support. Investment was instead concentrated largely in big state cooperatives and enterprises.

Despite these shortfalls, Sandinista agrarian reform—to improve living and working conditions of rural residents and modernize agriculture for the rural masses—was the most ambitious project in the country's history. But one of the undesirable legacies was the increased complexity of land-tenure and access arrangements which, in part, set the stage for a new round of land concentration with subsequent neoliberal governance, starting under President Chamorro. State land was (re)privatized and land under collective tenancy, like the Sandinista-era cooperatives, was parceled out—often on the cooperative members' own initiative. Many of the former landowners, from whom land had been expropriated, and who were often exiled in the United States during the 1980s, reclaimed their properties; additionally, the government promised land to ex-combatants (for details see Horton 1999). According to Martí i Puig and Baumeister (2017), the winners of this complicated land-tenure situation have been former owners (whose properties were nationalized), groups linked to the Sandinista National Liberation Front party (FSLN),[1] as well as to ex-leaders of the FSLN, who became part of the new agrarian bourgeoisie.

As shown by Baumeister and Fernández (2005), the Sandinista revolution and ensuing neoliberal governments have made no substantial improvements to the highly unequal distribution of lands between 1971 and 2001. During this period, the number of farms in Nicaragua almost doubled, as more lands were cleared on the agricultural frontier, but there was little to no change in the overall agrarian structure. With roughly 200 percent growth in population since 1960 (6.4 million in 2020), many farmers remain land-hungry. The middle sectors, owning farms between 35 and 350 hectares, benefited the most, but most farmers had as little access to lands in 2001 as before the revolution. Since 2001, the growth of medium-size farms has continued, although smallholders still have insufficient land access (Martí i Puig and Baumeister 2017).

Study region and methods

The study region (see Figure 6.1) is located in the hills of northern Nicaragua in the department of Madriz, which covers an area of 1,708 km^2 and has a population of 158,000 (INIDE 2011). Somoto, Madriz's capital, is located within a municipality of the same name, and this municipality constitutes our study region. Madriz is part of the so-called "dry corridor" (including the Pacific, north, and central regions of Nicaragua), which covers approximately 34 percent of the national territory and contains 80 percent of the country's population. The dry zone also coincides with those regions used most intensely for agricultural purposes (Segnestam 2009). The climate here is becoming drier and rains are less frequent, increasingly more intense, and more unpredictable (see World Bank 2013; Gourdji et al. 2015).

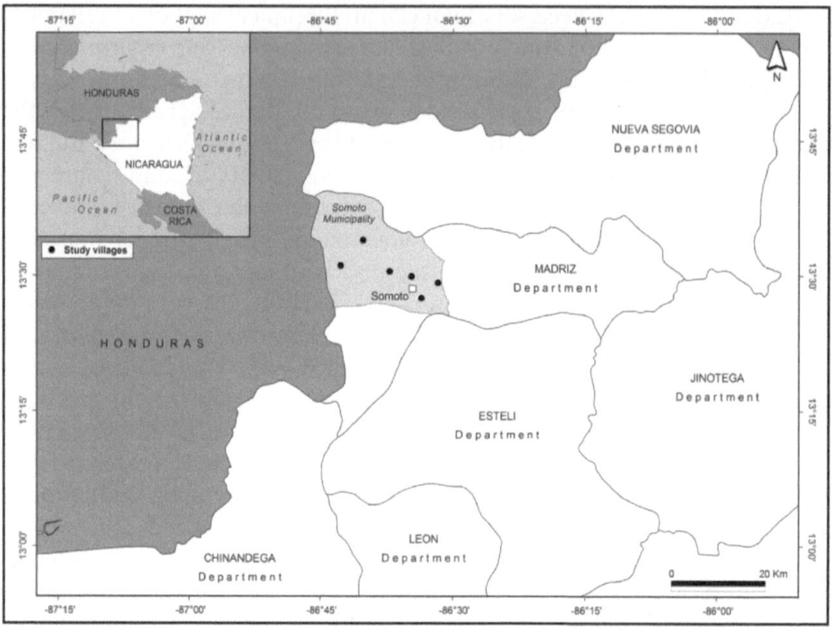

Figure 6.1 Study region.

Our research in Somoto is part of a larger project to examine migration and environment interactions in Nicaragua and Guatemala (see also Radel et al. 2018). We conducted research in 6 villages in the municipality of Somoto in 2014 and 2015: we applied a household survey to 120 families throughout the 6 villages in July and August 2014, and in June 2015 conducted 15 in-depth, semistructured interviews and 3 focus groups in 3 of the participating villages. Given the original research focus on international migration, we stratified the household survey sample (60 percent of households with a migrating member at the time of the interview or in the past; 40 percent of households with no migration experience) to ensure sufficient sampling of both groups. The survey instrument covered each household's demographic composition, agricultural activities, land access and tenure, international migration, and food security. In-depth interviews focused on themes of food security, agrarian change, migration, and gender. Focus groups, conducted using participatory techniques and activities, concentrated on producing detailed community histories on the same topics as the interviews.

Findings

Using survey data from all 120 households sampled in our study area of Somoto, we first provide an overview of our basic themes: land access and land tenure, agricultural production, and food (in)security. We then present qualitative findings from interviews and workshops in two of the six villages to illustrate how villagers perceive and describe their situation. We highlight these two villages as representing opposite ends in the experience of land tenure regimes prior to the country's neoliberal turn in the 1990s. These different histories shape villager relations and access to the Sandinista government. The community of La Cooperativa[2] was one of the well-known cooperatives during the 1980s, and during that time residents forged good relations with the revolutionary government. This community is located close to Somoto town and has a river nearby, which offers irrigation possibilities. The second community, Buena Vista, was never really "touched" by the revolution, as local landowners only had medium-sized farms and did not abandon their landholdings, which prevented expropriation of these holdings. Relations between Buena Vista residents and the Sandinista government have not been, and are not, good.

Land tenure and access

Around 50 percent of the sampled Somoto households own land (see Table 6.1), but the legal status of these parcels can vary. The other half of the households must rent or borrow land for cultivation.

Land access—a term that applies to land owned, borrowed, and rented—is low across our sample (see Figure 6.2), even though agriculture is the primary way people in the study area gain access to food and/or to income. Although some households are able to access much more, most households have access to less than 2 hectares—84 households, or 70.6 percent of sampled households—and 85 percent of the households have access to less than 3 hectares.

Agricultural production

Agricultural production is critical to household livelihoods in rural Somoto. Key crops for sampled families include maize and beans for household food consumption and sale, and cash crops like watermelons if irrigation is available. A minor crop, sorghum, is used for animal feed, and for food in desperate times. Some families with a little more land focus on milk and cheese production. Many families have at least one or two cows for subsistence production

Table 6.1 Land access and tenure (119 households*)

Households owning land	54.6% (65)
Households borrowing land	26.1% (31)
Households renting land	21.0% (25)

Source: Survey.

Note:
* For one of the 120 households, data on land access and tenure are not available.

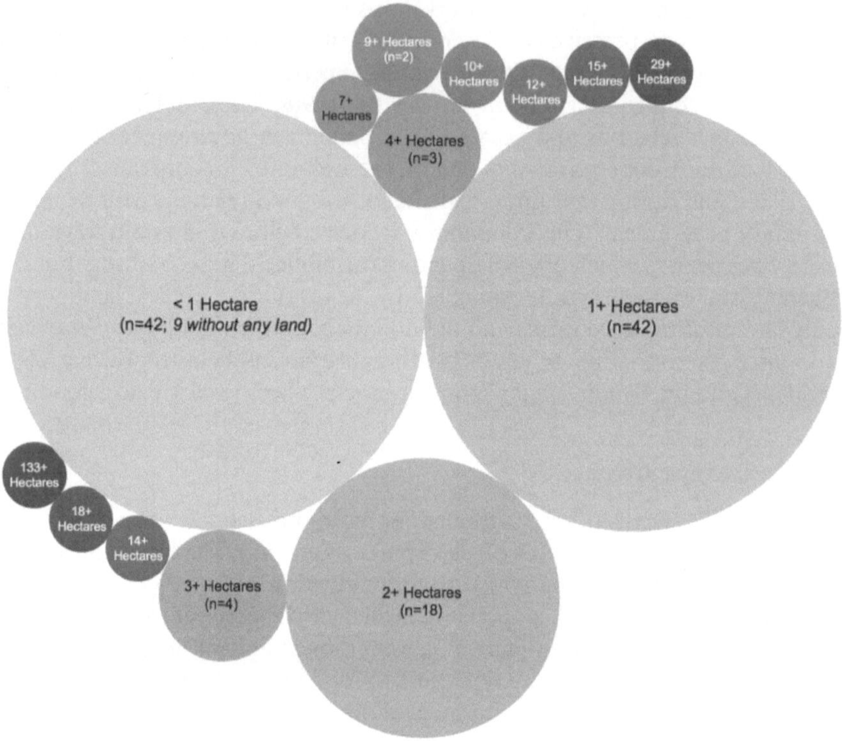

Figure 6.2 Land access in surveyed Somoto communities (119-household sample). Households are grouped into circles based on the amount of land accessed. The circle size represents the number of households in each land-access group. The circles are shaded light to dark as the amount of land accessed by households in that group increases.

Source: Survey.

of cheese, an important staple for the traditional meal, which includes rice and beans and a slice of cheese. Planting of crops occurs in two cultivation cycles. The first cycle of the calendar year is more important, as it takes advantage of the rainy season, but there is latent danger of irregular precipitation and dry spells. The dry period can be between five and eight months long, and normally only the few farmers who have access to irrigation infrastructure and water sources cultivate in the second cultivation cycle. As a result, more families plant maize during the first production cycle than during the second cycle. Our collected production data for the first annual cycle reflect the effects of a significant drought. In 2014, the region was hit hard by a drought that impacted most of Mesoamerica. Almost all interviewees who had planted maize during the first cycle (May to August) of that year lost everything (see Table 6.2). Given the limited land access and low yields, the semi-subsistence

Table 6.2 Agricultural production in 2013/2014, with drought (n = 119)

Crop	*Cycle (season)*	*Number of households (HH) planting*	*Mean ha planted/HH*	*Median ha planted/HH*	*Mean kg harvested/HH*	*Median kg harvested/HH*	*Mean yield kg/ha*	*Median yield kg/ha*
Maize	second (2013)	21	0.9	0.7	562	460	636	657
	first (2014)	31	0.9	0.7	0.2	0.00	0.2	0
Maize + Beans	second (2013)	30	1.1	1.2	460	460	426	376
	first (2014)	44	1.2	1.4	33	0.00	28.6	0
Beans	second (2013)	29	1.2	0.7	631	414	545	591
	first (2014)	14	0.7	0.7	2.3	0.00	3.3	0
Sorghum	second (2013)	11	0.9	0.7	776	322	853	460
	first (2014)	NA	NA	NA	NA	NA	NA	NA

Source: Survey.

Notes: The data here are only for households in the sample cultivating the crop in question as a monocrop, or in the case of Maize+Beans, as a dual planting. Additional sampled households reported mixed cropping, but the data do not allow for disaggregation. Yields reported here are calculated as the mean kg harvested / mean ha planted, and the median kg harvested / median ha planted, and are not survey-reported yields.

agriculture of these farmers barely provides a means of survival—there is little opportunity to save, invest, or build household wealth.

Food security/insecurity

In Somoto, food security and insecurity reflect production. Household-reported food availability over the year exhibits a relationship with agricultural production (see Figure 6.3). Maize is harvested in August and December, and in regular rainfall years the August harvest, the first, is the more important harvest (2014 was an aberrant year due to the drought, therefore the first harvest was much lower than expected). Beans are planted during both cycles, and often maize and beans are planted together. Food insecurity for households in the sample, as indicated by month-by-month reported food availability, is widespread in June and July, prior to the August harvest. A significant proportion of households, however, report food scarcities through the entire year. Most households were interviewed from May to June 2014, when food availability was dropping. At that time, 8 percent reported that at least one person in the family had skipped a meal over the previous four weeks, and 14 percent reported that at least one person had gone to bed hungry.

In sum, residents across the study communities have scarce access to land and are vulnerable to drought and subject to low yields. These factors, combined with the weak state support or safety net for small-scale farmers, result in residents' chronic struggle with seasonal hunger.

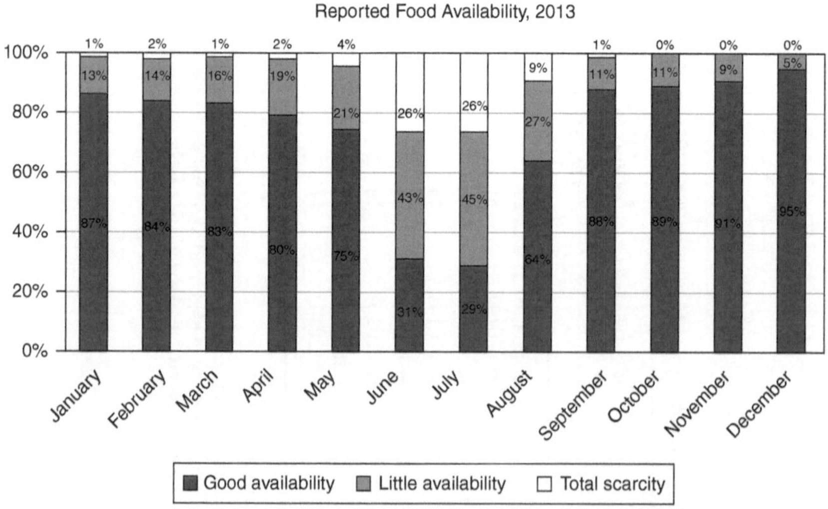

Figure 6.3 Household-reported food availability across the year (120 households). Source: Survey.

Two distinct communities, similar struggles for land and food security

Buena Vista is an agricultural community where, in the Sandinista years of the 1980s, medium to large landowners were unaffected by the land reform (and expropriations). These landowners retained their properties of approximately 1,000–1,500 *manzanas* (700–1050 ha). Many other people living in the region were displaced or migrated to Honduras in order to avoid violence or conscription during the ensuing US-sponsored Contra war against the Sandinista government (1979–90). Some people may have remained during this period and farmed land: research informants told the story that women stayed and cultivated in secret in order to survive.

Village livelihoods and cultivation practices have remained practically the same since the time of the Somoza dictatorship. Today, landless community members largely borrow land from the larger landholders, and many work as day laborers on the larger farms when opportunity arises. Maize and beans are produced mostly for subsistence, but a few families sell small surpluses. Commercial production of beans is around 20,000 *quintales* (1 *quintal* = 46 kg) for the community as a whole and is sold elsewhere. Residents reported that previously they needed only 3 *manzanas* (1 *manzana* = 0.7ha) of land in order to subsist, as maize and bean harvests were more abundant. Now they perceive needing more land to produce the same amount, due to the lack and irregularity of rainfall.

Today, residents access land in Buena Vista mostly through borrowing or renting the land from the community's larger landholders. These few families provide land in exchange for labor, part of the harvest, or leaving the remains of the harvest for the owner's cattle to eat. A resident explained:

> Here they go to the woman who owns that estate. She always gives them somewhere to plant, with the benefit that they return the land to her with pasture for her cattle. Here in Buena Vista there are many who plant on her land and do not rent land, because she lends it to them [with this requirement].

According to participants, these types of arrangements with landowners are not ideal for ensuring food security. Oftentimes, borrowing land means that community members are unable to produce enough food to last the year. As one workshop participant explained, this is because they are only able to cultivate once during the agricultural cycle, given the land rents or borrowing conditions.

Not having enough land to cultivate has negative consequences for food production, as peasants have to abandon their own food production to fulfill their obligation to work on the owner's plots. Therefore, families sometimes hire another person to work on the owner's plot to avoid abandoning the cultivation of their own food crops. One farmer from Buena Vista considered himself lucky to have found a steady arrangement to borrow land in exchange for work, but he still lamented the arrangement's sometimes-exploitative conditions:

> We have been lucky because we have not rented; the land has been given to us. We have already been working [the land] for about three years and we do not pay anything. There we have it [the land] … until they tell us "not anymore, go somewhere else." … They [the owners] also cultivate and we have to help them to work on their lands. It helps, but how should I put it, during planting time they take great advantage of us, and we have to stop working our own land and work on theirs. We have to look for someone to go there [to their land] in order to not stop working here [our own plot].

Another farmer explained:

> We suffer because sometimes we only harvest enough to feed our family because we do not work our own land. We work on rented land or we have to pay the landowner with our labor so that he will lend us a place where we can cultivate. And yet with these difficulties we have a lot of obligations.

One woman reported that landless peasants remember and miss the times when land was cheap. She related that some families with migrants in the United States have bought a *manzana* for US$2,000. Meanwhile, before, it sold at US$1,000 per *manzana* (according to an interview conducted on June 24, 2015). Migrants to Spain and people from Honduras have tried more recently to buy land in Buena Vista for as much as US$5,000 per *manzana*:

> Relatives in Spain and Honduras have wanted to buy [land] here, but he [my husband] says he will not sell. A young man from Honduras offered him $5,000 for a small piece [of land] because he wants to live here, but since he inherited it from his grandmother, he [my husband] cannot sell the land.

Looking at families' overall food security reveals that the conditions to satisfy daily food necessities are very fragile. Peasants, especially those who do not own land, find themselves in a difficult socioeconomic situation that does not guarantee their food security. For example, in Buena Vista, the landless have to engage in day labor, caring for cattle in order to buy food. Sometimes, the owner of the cattle may allow ranch-hand tenants to milk the cows and sell this milk to purchase foodstuffs. Securing access to food can be unpredictable. Another woman from Buena Vista recounts one occasion when they ran out of food because the rains destroyed the maize crop: "We had difficult times, but aid groups passed through and gave us provisions for four months, which helped us. I believe that without them we would have been found dead." Faced with the difficult situation of food shortages in Buena Vista, the Nicaraguan government has provided support for up to four months of "food packages" to families. However, these subsidies have been insufficient, and farmers have to supplement this governmental food aid with migration to El Salvador to earn the money necessary to buy inputs for the next agricultural season:

> Last year [2014] there was no harvest because it did not rain, but the government gave us a helping hand and gave us food packages. ...]It was rice, beans, oil, maize, salt, cereals. This helped us, because here in the countryside, this is what we eat. They helped us for six months. But we also had to go to El Salvador, because with that money we could buy 3 liters of pesticides and buy seeds and return to plant again.

La Cooperativa is an agricultural community cultivating maize, beans, tomatoes, peppers (*chiltoma*), squash (*pipián*), and watermelon. Residents of La Cooperativa express a firm pride in their history and celebrate their community's ability to organize for social and economic change. There is a strong sense of Sandinista identity in the community, stemming from its history as a political base community and agricultural cooperative. Before La Cooperativa became a cooperative in 1982, it was a cattle ranch. After the triumph of the Sandinista revolution in 1979, the lands of the ranch were distributed among peasant families as compensation for the damages of the civil war and, in 1984, the first houses were built. On the lands of the former cattle ranch the Sandinista government promoted one of the most important agricultural cooperatives in the region. The cooperative served as a source of jobs and thereby provided economic stability for peasant families.

With the end of the revolutionary period in 1990, the state dissolved the cooperative and cut off government subsidies, including extension services and credit. Cooperative members were forced to sell their land holdings in order to pay debts accrued after government supports were withdrawn. The cooperative ceased to exist, and many residents lost their land.

Today, families complement small-scale farming livelihoods with wage labor in neighboring tobacco fields. The daily wage of an agricultural day laborer is around 100 *córdobas* a day—equivalent to around US$3.75 at the time of research. Of the 120 families that live in the community, approximately 60 are landless and therefore have to rent or borrow land to cultivate. Depending on their incomes, families may be able to rent 1 to 3 *manzanas* of land.

In La Cooperativa we find different land access arrangements. Here more residents own their own land than do residents in Buena Vista. Some families own 10, 20, 30, 50, or even 80 *manzanas* of land. In this community, the cost of a *manzana* is reported at about US$3,000. In some cases, landless families look for several days, among landowners and relatives, for a day-labor job. Once they have gained the landowner's trust, they approach this owner to borrow a plot for cultivation. According to one participant, "some rent the *manzanas* for 7,000 *córdobas*, other [landowners] ask for 5,000 and others for 4,000 for an [agricultural] season of three months." There is some sense among the residents that landowners charge whatever they think can be justified. Community members explained that recently rents have increased. According to those interviewed, landowners who have observed the profitability of watermelon cultivation use this as an argument to increase rents. According to some respondents, landowners also have noted how people who have migrated to Spain have acquired land or livestock in La Cooperativa and are citing this as an explanation for higher rents.

The need to rent land in La Cooperativa reflects many residents' loss of land in previous years. During the 1990s, with Chamorro's new neoliberal government and left without government subsidies, extension services, and "cheap" credit, many families in financial need had to sell their lands. According to local residents, some families in better economic standing bought land from the needy. "Most of the people here had been given land, but they sold it because of their needs, and those who had a little more money appropriated the land."

Land inequality today in La Cooperativa presents a potential puzzle. How is it that some families had more resources to buy land after the Sandinista era than did others? Respondents suggested that with the disintegration of the cooperative and the subsequent sale of assets, like agricultural machinery and livestock, some former cooperative members had the possibility and means to acquire additional parcels. Some families also went through economic boom periods with watermelon cultivation, enabling them to buy land. As a result, some families now own more land than other families. During talks with community members, we found that among the ex-members of the cooperative there was an agreement not to sell the land to outsiders, which probably has caused the concentration of land in the hands of some of the ex-members.

A common struggle to access land and produce food

Remittances sent home by members in both communities under study are often important for land access. Residents migrate for short periods to Honduras, El Salvador, and Costa Rica; and for longer periods, or permanently, to Spain, France, Switzerland, and the United States (Radel et al. 2018; Carte, Radel, and Schmook 2019; Carte et al. 2019).

To get much-needed cash in La Cooperativa, families sometimes sell squash in neighboring villages in order to buy products such as rice, oil, or sugar to complement their own maize and bean harvests. In both communities, and in most cases, the family's agricultural production is barely sufficient to feed its members. Most families use part of the previous year's harvest as seeds for the next agricultural cycle and to pay the rent for the parcel they use.

Other families have had to rely on support from different organizations to improve their irrigation systems and to acquire seeds resistant to droughts and flooding, but this has been insufficient to improve/stabilize food production. In addition, prices for crops such as maize and beans fluctuate in the market. One farmer commented:

> We are agricultural communities and we are in a dry corridor, we do what is within our reach. Today we have to adapt to climate change. For example there are [now] some seeds that are more resistant, … and some organizations have given us a bit of seed and last year we had very little rain and [these seeds] yielded very little.

Pests and excessive rains have also led to crop losses in recent years. The difficult economic and agricultural-production situation has led families to experiment with different agricultural/productive techniques, but costs are very high. For example, a drip irrigation system costs 20,000 *córdobas* per *manzana* (approximately US$750 at the time of interviews). Peasant families in the two communities face elevated input costs, and they often find themselves in a dilemma over buying the irrigation system they want or other agricultural inputs. To these expenditures, landless families also need to add the cost of renting land. Margins are very narrow, even when farmers do manage to cover their input costs.

Conclusions

Food insecurity is a widespread problem in the study communities, as it is in many rural areas of Nicaragua. Food provision in the countryside mainly depends on households growing food for their own consumption, and the highly skewed distribution of agricultural lands results in limited and insecure land access for the rural poor. According to Martí i Puig and Baumeister (2017), while there was a slight decrease in the percentage of large landholdings, most farmers had little access to lands before the revolution and still find themselves in the same precarious situation today. The continued national policy orientation toward export agriculture, which has led to disinvestments in the small-scale food-production sector, is an additional impediment for achieving food security (HabibMintz 2004). Climate change, specifically increasing incidences of drought and increasingly irregular rainfalls, puts additional hardship on agricultural families.

Our two case-study communities, a former Sandinista cooperative and a remote village caught in feudal agrarian structures, each represent a different trajectory of land access and a unique set of responses to cope with land access and food insecurity. Together they exemplify the difficult situation in Nicaragua's countryside. In each of the two communities, the experience of food insecurity has a common cause: inequities in land assets and a lack of entitlements. The struggle of these residents highlights the historical agrarian continuities that lead to chronic food insecurity. While in the former Sandinista cooperative, with its still-strong ties to the government, residents can often rely on government support and international aid groups, in the other community residents have to rely more on the community's medium-to-large landowners and (to a minor degree) on the government and international aid groups. Additionally, remittances sent home to both communities are often important for land access, and therefore for food production. Despite a contemporary push to end hunger in rural spaces of Nicaragua, many people continue to experience hunger regularly as an outcome of power and asset inequalities that reflect contemporary agrarian policies and the persistence of hierarchal social relations.

Acknowledgements

We would like to thank our field and data-entry assistants, Leda Videa, Galileo Videa, Marjoryet Lazo, Jeyson Raudales, Navil Karelia García, Lisa Green, Juan Carlos Joo Chang, Rose Long, and Andres Zamudio. This chapter is based upon work supported by the National Science Foundation under Grant Number 1056811.

Notes

1 The FSLN (Frente Sandinista de Liberación Nacional/Sandinista National Liberation Front) is a socialist democratic political party in Nicaragua. The FSLN overthrew the Somoza dynasty and established a revolutionary government. Today the FSLN is again in power and one of the country's two leading parties (for more information see Horton 1999; Martí i Puig 2010).
2 Both the villages' names have been changed.

References

Bacon, Christopher M. 2015. "Food sovereignty, food security and fair trade: the case of an influential Nicaraguan smallholder cooperative." *Third World Quarterly* 36(3): 469–488, https://doi.org/10.1080/01436597.2015.1002991.

Bacon, Christopher M., William A. Sundstrom, María Eugenia Flores Gómez, V. Ernesto Méndez, Rica Santos, Barbara Goldoftas, and Ian Dougherty. 2014. "Explaining the 'hungry farmer paradox': Smallholders and fair trade cooperatives navigate seasonality and change in Nicaragua's corn and coffee markets." *Global Environmental Change* 25: 133–149, https://doi.org/10.1016/j.gloenvcha.2014.02.005.

Banco Central de Nicaragua (BCN). 2015. "Nicaragua en cifras [Nicaragua in figures]." Managua: BCN. Retrieved January 27, 2017, www.bcn.gob.ni/publicaciones/periodicidad/anual/nicaragua_cifras/nicaragua_cifras.pdf.

Baumeister, Eduardo, and Edgar Fernández. 2005. "*Análisis de la tenencia de la tierra en Nicaragua a partir del censo agropecuario 2001*." Managua: Ministerio de Agricultura Ganadena y Forestral (MAGFOR), Instituto Nicaragüense de Estadısticas y Censos (INEC), and the Food and Agriculture Organization of the United Nations (FAO).

Saturnino M., Borras, Jr., Jennifer C. Franco, and Sofía Monsalve Suárez. 2015. "Land and food sovereignty." *Third World Quarterly* 36(3): 600–617. https://doi.org/10.1080/01436597.2015.1029225.

Carte, L., Claudia Radel, and Birgit Schmook. 2019. "Subsistence migration: Smallholder food security and the maintenance of agriculture through mobility in Nicaragua." *Geographical Journal* 20(19) 185: 180–193, https://doi.org/10.1111/geoj.12287.

Carte, L., Birgit Schmook, Claudia Radel, and Richard Johnson. 2019. "The slow displacement of smallholder farming families: Land, hunger, and labor migration in Nicaragua and Guatemala." *Land* 8(6), 89, https://doi.org/10.3390/land8060089.

Food and Agriculture Organization of the United Nations (FAO). 1996. "The Rome Declaration on World Food Security," www.fao.org/docrep/003/w3613e/w3613e00.HTM

Food and Agriculture Organization of the United Nations (FAO). 2013. "Addressing the linkages between climate change and vulnerability to food insecurity. Testing a methodology in Nicaragua." www.fao.org/climatechange/59732/en/.

Food and Agriculture Organization of the United Nations (FAO). 2015. "Regional overview of food insecurity Latin America and the Caribbean." Retrieved January 27, 2017, www.fao.org/3/a-i4636e.pdf.

Godek, Wendy. 2015. "Challenges for food sovereignty policy making: The case of Nicaragua's Law 693." *Third World Quarterly* 36(3): 526–543, https://doi.org/10.1080/01436597.2015.1005437

Gourdji, Sharon, Peter Läderach, Armando Martínez Valle, Carlos Zelaya Martinez, and David B. Lobell. 2015. "Historical climate trends, deforestation, and maize and bean yields in Nicaragua." *Agricultural and Forest Meteorology* 200: 270–281, https://doi.org/10.1016/j.agrformet.2014.10.002.

HabibMintz, Nazia. 2004. "Food Insecurity in Nicaragua." Paper presented at the Annual ILASSA Conference on Latin America. Retrieved January 27, 2017, http://lanic.utexas.edu/project/etext/llilas/ilassa/2004/habibmintz.pdf.

Horton, Lynn. 1999. "*Peasant in arms: War and peace in the mountains of Nicaragua, 1979–1994.*" Athens, OH: Ohio Center for International Studies.

International Food Policy Research Institute (IFPRI). 2012. "2012 Global Hunger Index: The Challenge of Hunger: Ensuring sustainable food security under land, water, and energy stresses." Bonn/Washington, DC/Dublin: IFPRI, Concern Worldwide, and Welthungerhilfe and Green Scenery. Retrieved November 6, 2018, http://dx.doi.org/10.2499/9780896299429

International Fund for Agricultural Development (IFAD). 2012. "Republic of Nicaragua: Country strategic opportunities program." Retrieved March 3, 2017, https://webapps.ifad.org/members/eb/107/docs/EB-2012-107-R-11.pdf

Instituto Nacional de Información de Desarrollo (INIDE). 2011. "Anuario Estadístico 2011." Managua: Gobierno de Reconciliación y Unidad Nacional, Sistema Estadístico Nacional (SEN). Retrieved January 27, 2017, www.inide.gob.ni/bibliovirtual/Anuarios/ANUARIO11/anuario11.html#/37/zoomed.

Martí i Puig, Salvador. 2010. "The adaptation of the FSLN Daniel Ortega's leadership and democracy in Nicaragua." *Latin American Politics & Society* 52(4): 79–106, https://doi.org/10.1111/j.1548-2456.2010.00099.

Martí i Puig, Salvador, and Eduardo Baumeister. 2017. "Agrarian policies in Nicaragua: From revolution to the revival of agro-exports 1979–2015." *Journal of Agrarian Change* 17(2): 381–396, https://doi.org/10.1111/joac.12214.

Radel, Claudia, Birgit Schmook, Lindsay Carte, and Sofía Mardero. 2018. "Toward a political ecology of migration: Land, labor migration, and climate change in Northwestern Nicaragua." *World Development* 108: 263–273, https://doi.org/10.1016/j.worlddev.2017.04.023.

Sahley Caroline, Benjamin Crosby, David Nelson, and Lane Vanderslice. 2005. "The governance dimensions of food security in Nicaragua," USAID. Retrieved January 27, 2017, https://pdf.usaid.gov/pdf_docs/Pnade106.pdf.

Segnestam, Lisa. 2011. "Division of capitals—What role does it play for gender-differentiated vulnerability to drought in Nicaragua?" *Community Development* 40(1): 15–44, https://doi.org/10.1080/15575330903001562.

Trauger, Amy. 2014. "Toward a political geography of food sovereignty: Transforming territory, exchange and power in the liberal sovereign state." *Journal of Peasant Studies* 41(6): 1131–1152, https://doi.org/10.1080/03066150.2014.937339.

United Nations (UN). 2012. "Hunger: A national security threat." *UN Chronicle* 49(1 and 2). Retrieved January 27, 2017, https://unchronicle.un.org/article/hunger-national-security-threat.

World Bank. 2013. "Climate change impacts on water resources and adaptation in the rural water supply and sanitation sector in Nicaragua." Latin America and Caribbean Region Environment and Water Resources Occasional Paper Series. World Bank. Retrieved January 27, 2017, https://openknowledge.worldbank.org/handle/10986/16586.

World Bank and CIAT. 2015. "Climate-smart agriculture in Nicaragua. CSA Country Profiles for Africa, Asia, and Latin America and the Caribbean Series." Washington DC. Retrieved February 27, 2017, http://hdl.handle.net/10568/68244.

World Food Programme. 2018. "Country Programme Nicaragua 200434 (2013–2018)." Rome: World Food Programme Print 2018. Retrieved March 02, 2018, www1.wfp.org/operations/200434-nicaragua-country-programme-2013–2018.

7 Global water grabbing and food insecurity

Jampel Dell'Angelo, Maria Cristina Rulli, and Paolo D'Odorico

Introduction

Recent years have shown a particularly evident manifestation of the processes produced by the contemporary global agrarian transition: a large wave of transnational large-scale land acquisitions (LSLAs) has led several practitioners and scholars to speak of a new "global land rush." The strong societal implications associated with this reconfiguration of land ownership and transformation of the rural landscape of large areas of developing countries has raised political concerns on the unethical and unjust features of a dynamic that many denounce as a "land grabbing" (*Economist* 2009; FAO, IIED, and IFAD 2009). The multidimensional aspects of LSLAs have been scrutinized, and while there is a debate on the potentially positive impacts of transnational LSLAs, a large body of work has mainly pointed out its negative socio-environmental implications (see the special issues of *Canadian Journal of Development Studies* 2012; *Development and Change* 2013; *Globalisations* 2013; *Journal of Peasant Studies* 2012; *Third World Quarterly* 2013; and the meta-studies by Oberlack, Tejada, Messerli et al. 2016; Dell'Angelo et al. 2017a; 2017b).

A specific dimension of LSLAs most relevant for the development of this chapter is the hydrological one. As LSLAs are prevalently oriented toward intensive commercial farming production, the water-consumption implications are particularly important. The agricultural sector accounts for the largest share of freshwater resources for human appropriation (Oki and Kanae 2006; Hoekstra and Chapagain 2008) and global estimates show that water consumption for food production, including crops and livestock, can potentially account for approximately 85 percent of total societal water consumption (D'Odorico, Davis, Rosa et al. 2018). Securing water resources for agricultural production is a strategic priority of every country.

In this context different authors have highlighted that the main driver of transnational land acquisitions is water rather than land itself (Skinner and Cotula 2011; Allan, Kerletz, Sojamo et al. 2012; GRAIN, 2012; Mehta, Veldwisch, Franco et al., 2012; Woodhouse 2012; Franco, Mehta, Veldwisch et al., 2013). In LSLAs, companies (often backed by national governments) secure extensive amounts of land for intensive agricultural production for long period of times (often more than fifty years) through different types of concessions and leases. Such operations have a double benefit for the investors. They represent a potentially profitable investment for the private companies involved and at the same time they are coherent with strategic food security concerns of many governments in countries that depend on international trade in agricultural commodities.

It is not a coincidence that the largest spike in transnational LSLA and land investments happened soon after the food security and financial crises of 2008. For the countries that rely on virtual water imports though importation of agricultural commodities, land investments can be seen as a strategic national security response. This could be understood as a mechanism to increase food resilience and reduce vulnerability to food trade shocks (Marchand, Carr, Dell'Angelo et al. 2016; Seekell, Carr, Dell'Angelo et al. 2016). But the study of transnational land investments and the associated virtual water trade of agricultural commodities shows that in addition to the many negative social impacts, LSLAs substantially impact water resources in developing countries. Different studies have already quantified the amount of water that is appropriated through LSLAs' crop production (Rulli, Saviouri, and D'Odorico 2013; Breu, Bader, Messerli et al. 2016; Dell'Angelo, Rulli, and D'Odorico 2018) and assessed its potential impacts on the food security of the developing countries targeted by these investments (Rulli et al. 2013).

There are many definitions of water grabbing (see Mehta et al. 2012; Dell'Angelo et al. 2018). Water grabbing can refer to cases of illicit appropriation of agricultural water by the industrial sector (Wagle, Warghade, and Sathe 2012); it can refer to cases of contamination and freshwater resources pollution, which then make it unusable by the traditional users (Arduino, Colombo, Arcampo et al. 2012); or similar to cases of water abstraction upstream that affect less powerful groups downstream (Duvail, Médard, Hamerlynck et al. 2012). Water can be grabbed in the context of mining operations (Sosa, and Zwarteveen 2012), in the context of diversion of water flows for hydropower development (Islare et al. 2012), and as the result of agricultural intensification promoted by foreign direct investments (Bossio, Erkossa, Dile et al. 2012). These cases share a common denominator: they all describe situations of power imbalance in the appropriation of water resources, often in disregard of traditional users and their customary rights. Conceptually, Mehta and colleagues synthesize water grabbing as "a situation where powerful actors are able to take control of, or reallocate for their own benefits, water resources already used by local communities or feeding aquatic ecosystems on which their livelihoods are based" (2012: 197). Thus, this general definition can be applied to different socio-environmental and political processes of water appropriation.

In our work, we instead specifically refer to an operational definition of water grabbing that focuses on the hydrological dimension of LSLAs. We believe that the amounts of water that are potentially appropriated through land grabbing represent one of the most worrying threats to sustainable development and can undermine the achievement of several Sustainable Development Goals (SDGs) (Dell'Angelo et al. 2017b; D'Odorico et al. 2017). In particular, LSLAs can have a particularly strong impact on global food security (Rulli and D'Odorico 2014). The observation of a global phenomenon of transnational virtual water appropriation through LSLAs leads us to speak of a "Global Water Grabbing Syndrome" (Dell'Angelo et al. 2018). We are particularly concerned about the impacts of this phenomenon on countries that have high levels of malnutrition and deal with structural food security problems. In this chapter we illustrate and apply an analytical framework that we previously developed to characterize the relationship between LSLAs, virtual water appropriation, and food security (see Dell'Angelo et al. 2018).

The global water grabbing syndrome

Schellnhuber, Block, Cassel-Gintz and colleagues (1997) introduced the term "syndrome" as a concept useful to represent widespread globalized phenomena, both in the etymological ancient Greek sense as "things running together" and in the clinical meaning of a phenomenon that manifests clear symptoms. It is in this dual sense that Dell'Angelo and colleagues (2018) describe the hydrological processes associated with the contemporary escalation of large-scale land investments as a "global water-grabbing syndrome." Assessing global water grabbing implies both an analytical challenge as well as normative one. There is a debate around the issue that not all LSLAs should be automatically tagged as land grabs because many of them could be possibly conducted following ethical standards and could benefit local populations through development opportunities. How to identify when an LSLA is a "land grab" is not completely clear, different standards such as the FAO "voluntary guidelines on the responsible governance of tenure of land fisheries and forests in the context of national food security" provide advice on land investments. More directly, the International Land Coalition stated in 2011 Tirana Declaration the ethical, moral, and juridical conditions under which land investments constitute grabs (ILC 2011). These kinds of documents, however, do not provide an operational framework that can be applied systematically.

Considering water grabbing as the hydrological dimension of land grabbing implies that there is the same normative constraint related to the conditions necessary to determine when LSLAs should be considered a form of grabbing. Here we build on the framework Dell'Angelo and colleagues (2018) developed, which for the rest of the chapter we refer to as the "global water grabbing (GWG) framework." This framework proposes overcoming the normative question about the "grabbing" connotation by including two indicators, a biophysical one and an ethical one.

Assessing the likelihood of LSLAs-related (blue) water grabs

The GWG framework studies the hydrological and human-rights traits of land grabbing, addressing both the biophysical and the ethical dimensions. To delineate water grabbing in relation to large-scale land acquisitions, the GWG framework first differentiates between green-water and blue-water grabbing. The framework then deals with the normative/value-laden feature of the term "grabbing" by invoking the human rights approach to food. Under Article 11 of the International Covenant on Economic, Social and Cultural Rights (UN 1966) every state is obliged to ensure that all its people have access to the minimum amount of nutritionally adequate and safe food in order to be free from hunger (De Schutter 2009:2). Article 25 of the Universal Declaration of Human Rights declares that: "(1) Everyone has the right to a standard of living adequate for the health and well-being of himself and of his family, including food, clothing, housing and medical care and necessary social services" (D'Odorico, Carr, Davis et al. 2019).

Rather than referring to the United Nations human right to safe drinking water, the GWG framework invokes the right to food based on the motivation that most human appropriation of freshwater resources is for agriculture (>80%) and not for drinking (<1%) (Hoekstra and Chapagain 2008; D'Odorico et al. 2018), and crop production is mostly constrained by physical and economic water scarcity (e.g., Mueller, Gerber, Johnson et al. 2012).

This rationale is based on the fact that the nexus between food production and water is the central element that determines both the biophysical consequences of LSLAs and the ethical limits beyond which these implications should be negatively assessed. Therefore "water grabbing" potentially happens in those countries that have scarce water resources and do not exploit their full potential for agricultural production to satisfy their food security and nutritional need, while foreign and/or private interests impose their own commercial priorities despite the existence of an endemic condition of food scarcity. The normative dimension of this assessment refers to the UN article 11 (UN 1966). The acquisition of land and water resources is defined as unethical because those resources should instead be used to reduce malnourishment. If these conditions are violated then, according to the GWG framework, the hydrological appropriations associated with LSLAs are likely to constitute a "water grab." In sum, we define the high likelihood of blue-water grabbing will occur when LSLAs happen in a country that has high levels of malnutrition and high levels of water scarcity. In other words, the likelihood that the LSLAs happening in a country with high levels of water scarcity and high levels of malnourishment are a blue-water grab is, according to this framework, high.

The different analytical phases to apply the GWG framework are the following: The first step of this diagnostic approach is to identify the LSLAs to be included in the assessment. The framework uses data from The Land Matrix (TLM 2016). TLM (2016) reports information on size, location, negotiation status, and cultivated crops of each land deal. Up to December 2016, forestry and agriculture land deals under contract covered an area of approximately 40 million hectares where about 11 percent of the hectares (4.5 million ha) was under production. The assessment produced by Dell'Angelo and colleagues (2018) included 1,312 land deals with concluded contracts; 780 of these deals were in operation.

The second step is to discriminate between the blue versus green hydrological traits of the LSLAs. The GWG framework first determines whether LSLAs in every specific country prevalently appropriates green or blue water. While green water coincides with land appropriation, the assessment of blue-water appropriation is more complex because it happens only if the land is irrigated, which is the case when there are relatively high ratios, between the potential blue water use by crops (in hydrologically optimal conditions for crop productivity) and the total blue and green water use by crops optimally irrigated. In the assessment we only look at blue water because we are specifically interested in the grabbing of water resources that can be employed for irrigation. This is important: whereas green water is attached to the land and is acquired with it, blue water is withdrawn from water bodies and delivered to farmlands by irrigation systems, canals, and conduits. In areas where blue-water resources are scarce, competition arises over the appropriation of these resources. Therefore, blue-water appropriation by land investors who have the capacity to invest in hydraulic infrastructures may lead to dispossession of local communities of their water and/or exclusion of other farmers from future access to these water resources.

The third step relates to the biophysical indicator of water scarcity. Assessing water scarcity at the country level is complex. There are two main definitions of water scarcity, a socio-economic one and a physical definition. The first definition refers to countries where, despite a relative abundance of water resources, the human water demands are not satisfied because of economic, institutional, and technological constraints. The second definition refers to the low availability of water in nature in relation to the potential water requirements of the country. Different indicators have been used to describe one or more dimensions of water scarcity (Falkenmark, Lundquist, Widstrand et al. 1989; Shiklomanov 1991; Raskin, Gleick, Kirshen et al. 1997; Seckler, Amarasinghe, Molden et al. 1998; Alcamo, Henrichs, Thomas, Rosch et al. 2000; Vorosmarty, Green, Salisbury et al. 2000; Sullivan, Meigh, Giacomello 2003). The GWG framework uses Brauman and colleagues' (2016) depletion metric to assess water scarcity. Brauman (2016) defines water scarcity at basin level as the fraction of available renewable water consumptively used by anthropic activities, and they account also for seasonal variations. The GWG framework classifies mid-to-high water scarcity in a specific country if more than 15 percent of their area shows scarcity higher than 75 percent (seasonal and dry year included). It could be counter-argued that in the cases where economic water scarcity prevails, local populations do not use water anyway because of a lack of infrastructure. This criticism does not hold though: it is fundamental to also consider the rights of these populations to potential future water use.

The fourth step is to integrate data on the prevalence of undernourishment as the ethical indicator of the assessment. According to the framework, malnutrition is a fundamental dimension in the identification of water grabbing. It accounts for the normative/justice characteristic of the term "grabbing" and is immediately associated with water. The linkage is that in countries with malnourishment, land and water resources could (and we state, *should*) be used to satisfy local food needs (Rulli and D'Odorico 2014). This is directly connected with the right to food (DeSchutter 2009, 2011). It has already been pointed out by different international organizations that when national food security is under threat, as in cases of severe droughts, national trade policies should control the capacity of foreign investors to export food commodities (von Braun and Meinzen-Dick 2009). It is clear that blue-water appropriations in areas where water resources are already depleted will compete with the water needs of local populations. This explains how the combination of an indicator of malnutrition with a hydrological indicator of water scarcity highlights the nexus existing between blue water and food production in water-scarce regions. Approximately 92 percent of the blue-water footprint of humanity is for food production, agriculture, and livestock, and the largest share of this is for food (Richter 2014; D'Odorico et al. 2018). The GWG framework uses the Food and Agriculture Organization of the United Nations (FAO) 2015 data on prevalence of undernourishment (PoU). This database classifies countries as having low or moderately low food insecurity when less than 15 percent of their population is undernourished, and high or moderately high when the prevalence of undernourishment in the country exceeds 15 percent (FAO, IFAD and WFP 2015). The same 15 percent threshold to classify food secure and food insecure countries is used in the assessment. While PoU data are still widely used among international organizations, in particular by the FAO, and still constitute the most recognized measure for international analytical comparisons, there are several limitations. Aspects such as the focus on caloric needs for a sedentary life, the lack of attention to shorter-term hunger issues, and a lack of consideration for other dimensions beyond caloric intake such as quality of food make this measure narrow (Lappé, Clapp, Anderson et al. 2013). The PoU data are instrumental for studies such as the one we present here, however the recent development of multidimensional indicators that better account for the complexity of food insecurity—such as the self-reported experiential measures of food insecurity provided by the 'Food Insecurity Experience Scale' (FIES)—is very promising and represents a fundamental complement to the conventional indicators (Smith, Rabbit, Coleman-Jensen et al. 2017).

The last step of the GWG framework is to assess the likelihood that the LSLAs in a certain country are leading to blue-water grabbing (see Figure 7.1). Based on the previous assumptions and phases of operationalization, the framework identifies countries based on the combination of the two 15 percent

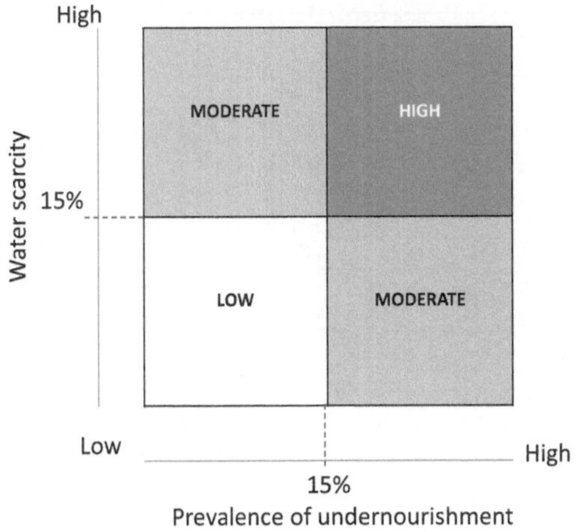

Figure 7.1 Likelihood of blue-water grabbing associated with LSLAs from Dell'Angelo et al., 2018.

thresholds for the malnutrition and water scarcity indicators (it is a coincidence that 15 percent is the threshold that makes sense for both indicators).

Therefore, the GWG framework identifies target countries where LSLAs have a high likelihood to be water grabs (see Figure 7.2), based on two conditions: (1) a level of undernourishment higher than 15 percent and (2) blue-water scarcity levels higher than 15 percent. Conversely, when the two conditions are not met the likelihood is low, while if only one of the two conditions is met the likelihood is medium.

Identifying the potential impact of water grabbing on food security

The acquisition of land and land-based resources, including water, may affect food security both at local scales and beyond (D'Odorico and Rulli 2013). In fact, the acquired land and water could be used to produce food for local communities. Large-scale land acquisitions contribute to a transition from subsistence farming to large-scale production by agribusiness corporations, often for export markets (DeSchutter 2011). Despite some exceptions (Lisk 2013) the food produced in the acquired land is often sold on the international market at prices that are not accessible to the local communities (Chinsinga, Chasukwa, Zuka et al. 2013; Lavers 2012).

The counterargument has often been that relatively large yield gaps exist in regions of the developing world where investments in modern agricultural technologies have been lacking, thereby preventing the development of adequate irrigation systems. In this light, LSLAs could bring the financial resources and knowledge needed to utilize land and water resources more

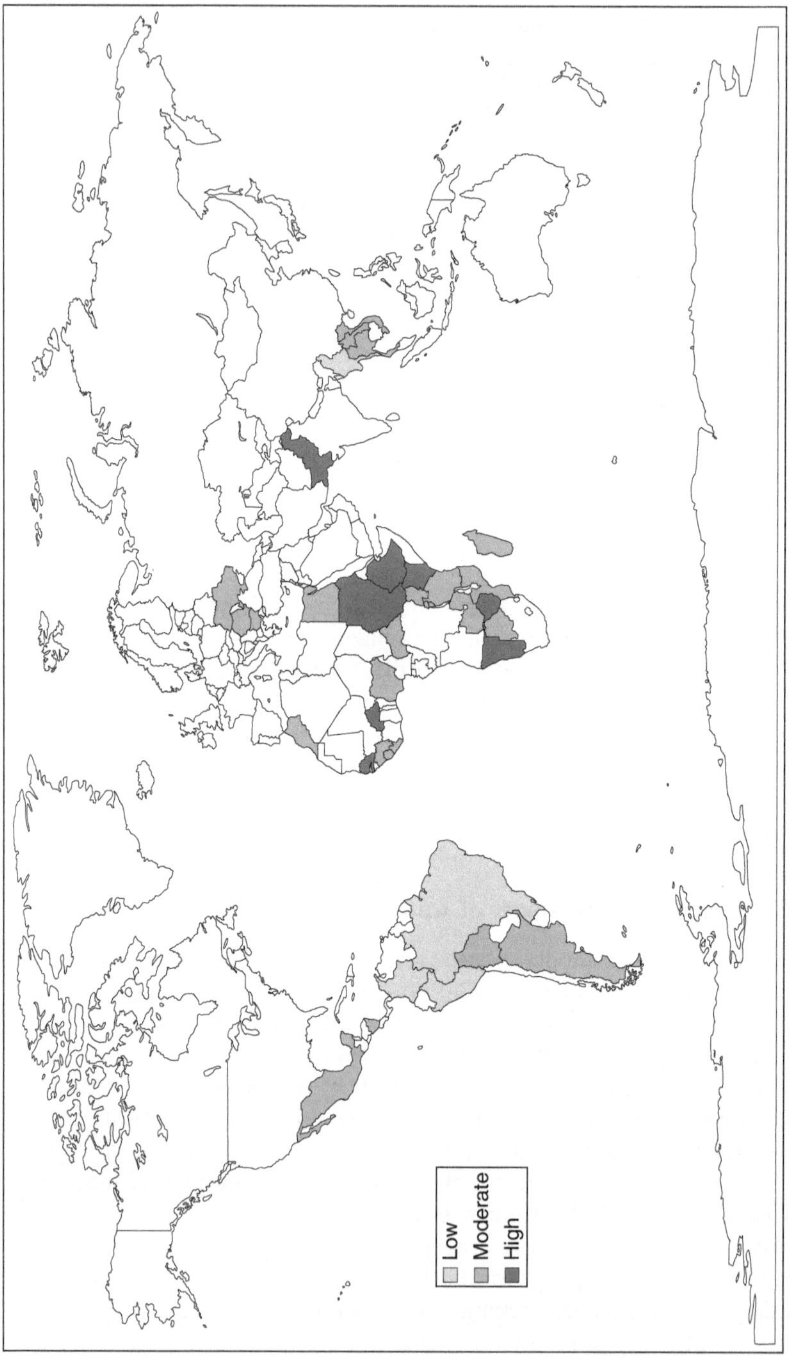

Figure 7.2 Map of the global distribution of the likelihood of blue-water grabbing associated with LSLAs based on GWG framework from Dell'Angelo et al., 2018.

effectively and consequently improve crop yields, food production, and food security (e.g., Deininger and Byerlee 2011a; Chakrabarti and Da Silva 2012). Promoters of LSLAs have also argued that in many cases investors target land that was not previously used by local communities for agriculture; therefore, no direct competition should arise between LSLAs and local food security. But no land is ever completely not utilized, and in many cases local communities still rely on it for firewood, bushmeat, grazing, honey production, or other ecosystem services (D'Odorico, Rulli, Dell'Angelo et al. 2017). In many cases land investors do not even put the land under production (TLM 2017), which means that LSLAs do not lead to an increase in food production, employment, technology transfer, or access to markets, but local communities may lose access to the land and its resources (D'Odorico et al. 2017).

The impact of land and water acquisitions on local food security is often difficult to evaluate (Rulli and D'Odorico 2014); although LSLAs can increase food production, local communities may not benefit from these investments if the production is for the export markets and profits are not shared with the host countries. Thus, the impact of land and water grabbing on food and water security will depend on three conditions: (1) the extent to which local populations are affected by undernourishments and the violation of human rights to food; (2) whether the acquired land and water resources were previously used by agriculture to produce food; and (3) the extent to which the acquired land is put under production and its products remain available to local communities.

Beyond the likelihood that LSLAs in certain countries represent water grabs, what is the impact on food security there? To answer this question, we here combine the GWG framework with the results of the assessment produced by Rulli and D'Odorico (2014) on the food that is appropriated through LSLAs (see their paper for materials and methods). A quantitative assessment of the potential impact of LSLAs on food availability was presented by Rulli and D'Odorico (2014), who calculated the amount of food that could be produced in the acquired land, with the crop yields reported for the year 2000 as well as at yield gap closure.

Combining the GWG framework with the analysis conducted by Rulli and D'Odorico (2014) we estimate the number of people that could be fed in a selection of LSLAs target (Table 7.1). This assessment, rather than being understood as a proposition and endorsement for a certain type of rural development, should be understood more as a quantitative representation of the food-production capacity that transnational LSLAs can potentially undermine.

Table 7.1 shows the people who could be fed with the food that is potentially produced by LSLAs, considering both an ideal balanced diet (column 5) and the average diet for each country (column 6). These numbers produce a good representation of the magnitude of the potential "food grabbing" associated with LSLAs in countries that have different degrees of likelihood of water grabbing.

Table 7.1 Application of the GWG framework to identify the amount of food that could be produced in LSLAs in target countries by adopting two different diet scenarios

Country	*Prevalence of Undernourishment SOFI 2015 – 15%*	*Water scarcity index 15%*	*Likelihood of grabbing*	*Millions of people that could be fed*	
				Ideal diet	*Aver. diet*
Ethiopia	32	21%	High	20.1	24.7
Zimbabwe	33.4	48%	High	1.5	2.3
Guatemala	15.6	0%	Moderate	0.2	0.3
Liberia	31.9	0%	Moderate	6.7	14.1
Madagascar	33	12%	Moderate	1.8	2.9
Mozambique	25.3	5%	Moderate	4.5	7.2
Sierra Leone	22.3	0%	Moderate	12.9	17.1
Tanzania	32.1	4%	Moderate	5.2	7.6
Uganda	25.5	1%	Moderate	0.2	0.2
Argentina	<5.0	16%	Moderate	4.7	5.2
Nigeria	7	16%	Moderate	2.3	5.8
Morocco	<5.0	84%	Moderate	0.6	0.6

Discussion and conclusions

This chapter discusses the essential relationship between food security and water grabbing, and it proposes a numerical assessment that can be used to develop a better understanding of the magnitude and extension of the phenomenon. Nevertheless, the questions and concerns about LSLAs extend beyond how much water, land, and food are appropriated and exported overseas at the expense of local populations and how much food could be produced to satisfy local food demands, but is not. The central problem is much broader, and it calls into question some fundamental processes of the global agrarian transition.

While this assessment shows that there are millions of people who could be fed in countries that have high and medium likelihood of water grabbing associated with LSLAs, it does not suggest that large-scale agricultural production, if used for the right purpose, could have a beneficial impact. The overarching question about the type of agricultural development that is fairer, more environmentally sound, and has the strongest potential to reduce poverty and malnutrition and enhance wellbeing, still remains.

To address this question, it is not enough to discuss only the negative, actual, or potential outcomes of LSLAs. It is instead fundamental to tackle the core of the debate about the different rural development trajectories and economic models that are discussed pertaining to the future of developing countries (Dell'Angelo et al. 2017a). One of the central issues in this debate is about the implications of a capitalist restructuring of global agriculture versus the role of small-scale farming and traditional systems. This debate could be synthetically described as a clash of opposing models with very different presumptions, implications, and conflicting priorities (Rosset 2003). LSLAs are part of a vision of development to advocate for a rural transition in developing countries; that vision carries technological improvements, modernization of agricultural systems, development of infrastructure, employment of biotechnologies, commercialization, commoditization and liberalization of agricultural products, integration in global markets, enhancement of exporting capacity, and of the capacity to attract foreign corporate capital and financial investments (Deininger, Byerlee, Lindsey et al. 2011; Byerlee and Deininger 2013; Arezki, Deininger, and Selod 2013; Chakrabarti and Da Silva 2012; Hallam 2011).

On the opposite side there is the defense of many of the benefits of small-scale farming systems. For example, it has been described how small-scale farmers are able to amply satisfy the world food demands and how small farms have higher crop diversity than large farms (IAASTD 2009; Ricciardi, Ramankutty, Mehrabi et al. 2018). Moreover, traditional rural systems have strong socio-economic implications: for example, they allow for livelihood diversification strategies and represent a positive force against inequality and resource distribution (Scheidel et al. 2013; Ravera et al. 2014). Often small-scale farming happens in a context of common property systems and community dynamics. Another example of important traditional rural systems of production that escape the capitalist transformation are the pastoralist ones, which are the result of thousands of years of socio-ecological adaptation and are practiced worldwide by between 200 and 500 million people (McGahey et al. 2014; Dell'Angelo et al. 2017b).

This clash of models is well represented by the contrast between the dominant corporate neoliberal agricultural vision and the ones proposed by international farmers' and peasants' social movements such as the Via Campesina (Rosset 2003). The Via Campesina, one of the strongest global agrarian movements, explicitly fights for food sovereignty, peasants' rights, and environmental justice against agribusiness, transnational companies, and agricultural commercial speculation. From Via Campesina's influential perspective, food sovereignty is a key political tenet that goes well beyond the concept of food security. Agricultural and food production is not only about the satisfaction of the nutritional needs of each human being, but it is also about the conditions of life and existence that are associated with the production of that food.

The "food question" is an extension of the "land question." Water grabbing, independently from the metrics and definitions that are used to identify it, is ultimately the strongest biophysical manifestation of this question. It is fundamental to understand that water is ultimately the critical factor for rural justice, and that the issue of hunger and food security needs to be addressed not only from a technical agricultural point of view but also as part of a dynamic that involves ethics, struggle, power dynamics, and politics.

Last, global quantitative assessments such as the ones that we present here provide valuable information, but they need to be completed with a thorough understanding of the conditions on the ground and a discussion of the political implications that are associated with the emerging trends in rural transitions. Water and land grabbing and food insecurity are associated with stories of pain, suffering, struggle, injustice, and resistance that studies such as the one presented in this chapter do not describe. The social implications of LSLAs have been thoroughly illustrated by an emerging body of scholarship on land grabbing based on field work and ethnographic research though these publications (see special issues in: *Canadian Journal of Development Studies* 2012; *Development and Change* 2013; *Globalisations* 2013; *The Journal of Peasant Studies* 2012; and *Third World Quarterly* 2013). This literature shows that the large majority of the cases of land grabbing happened with clear dynamics of coercion in the contexts of small-scale farming, common property regimes, or multiple-ownership claims (Dell'Angelo et al. 2017). A fundamental question that has been raised considering the subjects that are impacted by these dynamics is "what are the political reactions from below?" (Hall et al. 2015). A collection of articles in the special issue on global land grabbing and political reactions "from below" address this question looking at the different responses that go from direct resistance to different types of forms of mobilization and counter-mobilization (*Journal of Peasant Studies* 2015). It becomes clear that there is a dialectic relationship between micro-processes of resistance and larger-scale processes associated with land and water grabbing that is fundamental to account for in order to understand the multi-scalar dimensions of food, land and water struggles (Temper 2018).

Acknowledgments

This work was partially supported by the National Socio-Environmental Synthesis Center (SESYNC) under funding received from the National Science Foundation DBI-1052875. J.D'A, M.C.R and P.D'O are part of the NEWAVE, Innovative Training Network (ITN), Marie Skłodowska-Curie Action (MSCA) - grant agreement No. 861509.

References

Alcamo, Joseph, Thomas Henrichs, and Thomas Rosch, 2000. "World Water in 2025: Global Modeling and Scenario Analysis." In Rijsberman (ed.), *World Water Scenarios Analyses*. Marseille: World Water Council.

Allan, John Anthony, Martin Kerlertz, Suvi Sojamo, and Jeroen Warner. 2012. *Handbook of Land and Water Grabs in Africa: Foreign Direct Investment and Water and Food Security*. London: Routledge.

Arduino, Serena, Giorgio Colombo, Ofelia Maria Ocampo, and Luca Panzeri. 2012. "Contamination of Community Potable Water from Land Grabbing: A Case Study from Rural Tanzania." *Water Alternatives* 5(2): 344.

Arezki, Rabah, Klaus Deininger, and Harris Selod. 2013. "What Drives the Global 'Land Rush'?" *The World Bank Economic Review* 29(2): 207–233.

Bossio, Deborah, Teklu Erkossa, Yihun Dile, Matthew McCartney, Franziska Killiches, and Holger Hoff. 2012. "Water Implications of Foreign Direct Investment in Ethiopia's agricultural sector." *Water Alternatives* 5(2): 223.

Brauman, Kate, Brian Richter, Sandra Postel, Marcus Malsy, and Martina Flörke. 2016. "Water Depletion: An Improved Metric for Incorporating Seasonal and Dry-Year Water Scarcity into Water Risk Assessments." *Elementa* 4(1): www.elementascience.org/articles/10.12952/journal.elementa.000083/.

Breu, Thomas, Christoph Bader, Peter Messerli, Andreas Heinimann, Stephan Rist, and Sandra Eckert. 2016. "Large-scale land acquisition and its effects on the water balance in investor and host countries." *PLoS One* 11(3): e0150901.

Byerlee, Derek and Klaus Deininger. 2013. "Growing Resource Scarcity and Global Farmland Investment." *Annual Review of Resource Economics* 5(1): 13–34.

Canadian Journal of Development Studies. 2012. Special Issue on Land Grabbing in Latin America. *Canadian Journal of Development Studies* 33(4).

Chakrabarti, Suma, and Jose Graziano Da Silva. 2012. "Hungry for Investment: The Private Sector Can Drive Agricultural Development in Countries That Need It Most." *Wall Street Journal*, September 6.

Chinsinga, Blessings, Michael Chasukwa, and Sane Pashane Zuka. 2013. "The Political Economy of Land Grabs in Malawi: Investigating the Contribution of Limphasa Sugar Corporation to Rural Development." *Journal of Agricultural and Environmental Ethics* 26(6): 1065–1084.

D'Odorico, Paola, and Maria Cristina Rulli. 2013. "The Fourth Food Revolution," *Nature Geoscience* 6(6): 417–418. DOI: 10.1038/ngeo1842.

D'Odorico, Paola, Maria Cristina Rulli, Jampel Dell'Angelo, and Kyle Frankel Davis. 2017. "New Frontiers of Land and Water Commodification: Socioenviro nmental Controversies of Large-Scale Land Acquisitions." *Land Degradation & Development*. DOI: 10.1002/ldr.2750.

D'Odorico, Paolo, Kyle Frankel Davis Lorenzo Rosa, Joel A. Carr, Davide Chiarelli, Jampel Dell'Angelo et al. 2018. "The Global Food-Energy-Water Nexus." *Reviews of Geophysics* 56(3): 456–531. DOI: 10.1029/2017RG000591.

D'Odorico, Paolo, Joel A. Carr, Kyle Frankel Davis, Jampel Dell'Angelo, and David Seekell. 2019. "Food Inequality, Injustice, and Rights." *Bioscience* 69(3): 180–190.

De Schutter, Olivier. 2009. "Large-Scale Land Acquisitions and Leases: A Set of Core Principles and Measures to Address the Human Rights Challenge." Briefing Note. June 11, 2009. Geneva: UN Office of the High Commissioner for Human Rights.

De Schutter, Olivier, 2011. "How Not to Think of Land-Grabbing: Three Critiques of Large-Scale Investments in Farmland." *Journal of Peasant Studies* 38(2): 249–279. DOI: 10.1080/03066150.2011.559008.

Deininger, Klaus, and Derek Byerlee. 2011. *Rising Global Interest in Farmland: Can It Yield Sustainable and Equitable Benefits?* Washington, DC: The World Bank.

Deininger, Klaus, and Derek Byerlee. 2011a. *The Rise of Large Farms in Land Abundant Countries: Do They Have a Future?* Washington, DC: The World Bank.

Deininger, Klaus, and Derek Byerlee. 2011b. *Rising Global Interest in Farmland: Can It Yield Sustainable and Equitable Benefits?* Washington, DC: The World Bank.

Dell'Angelo, Jampel, Paolo D'Odorico, and Maria Cristina Rulli. 2017a. "Threats to Sustainable Development Posed by Land and Water Grabbing." *Current Opinion in Environmental Sustainability* 26: 120–128.

Dell'Angelo, Jampel, Paolo D'Odorico, Maria Cristina Rulli, and Philippe Marchand. 2017b. "The Tragedy of the Grabbed Commons: Coercion and Dispossession in the Global Land Rush." *World Development* 92: 1–12.

Dell'Angelo, Jampel, Maria Cristina Rulli, and Paolo D'Odorico. 2018. "The Global Water-Grabbing Syndrome." *Ecological Economics* 143: 276–285.

Development and Change. 2013. Special Issue. Governing the Global Land Grab: The Role of the State in the Rush for Land. *Development and Change* 44(2).

Duvail, Stephanie, Claire Médard, Olivier Hamerlynck, and Dorothy Wanja Nyingi. 2012. "Land and Water Grabbing in an East African Coastal Wetland: The Case of the Tana Delta." *Water Alternatives* 5(2): 322.

Economist. 2009. "Buying Farmland Abroad: Outsourcing's Third Wave." Retrieved from: www.economist.com/node/13692889.

FAO, IIED, and IFAD. 2009. *Land Grab of Development Opportunity? Agricultural Investment and International Land Deals in Africa*. Rome: FAO.

Falkenmark, Malin, Jan Lundquist, and Carl Widstrand. 1989. "Macro-Scale Water Scarcity Requires Micro-Scale Approaches: Aspects of Vulnerability in Semi-Arid Development." *Natural Resource Forum* 13(4): 258–267.

FAO, IFAD, and WFP. 2015. The State of Food Insecurity in the World 2015. Meeting the 2015 International Hunger Targets: Taking Stock of Uneven Progress. Rome: FAO.

Franco, Jennifer, Lyla Mehta, Gert Jan Veldwisch. 2013. "The Global Politics of Water Grabbing." *Third World Quarterly* 34(9): 1651–1675.

Globalisations. 2013. Special Issue: Land-Grabbing and Global Governance." *Globalisations* 10(1).

GRAIN, 2012. "Squeezing Africa Dry: Behind Every Land Grab is a Water Grab. Barcelona." www.grain.org/article/entries/4516-squeezing-africa-dry-behind-every-land-grab-is-a-water-grab. Retrieved June 2016.

Hall, Ruth, Marc Edelman, Saturnino Borras Jr, Ian Scoones, Ben White, and Wendy Wolford. 2015. "Resistance, Acquiescence or Incorporation? An Introduction to Land Grabbing and Political Reactions 'From Below'." *Journal of Peasant Studies* 42(3–4): 467–488.

Hallam, David. 2011. "International Investment in Developing Country Agriculture—Issues and Challenges." *Food Security* 3(1): 91–98.

Hoekstra, Arjen, and Ashok Chapagain. 2008. *Globalization of Water*. New York: John Wiley and Sons. DOI: 10.1002/047147844X.wr51.

Holmén, Hans. 2015. "Is Land Grabbing Always What It Is Supposed to Be? Large-Scale Land Investments in Sub-Saharan Africa." *Development Policy Review* 33(4): 457–478.

International Land Coalition (ILC). 2011. *Global Assembly 2011*. Tirana, Albania: ILC.

Islar, Mine. 2012. "Privatised Hydropower Development in Turkey: A Case of Water Grabbing?" *Water Alternatives* 5(2): 376–391.

Journal of Peasant Studies. 2012. Special Issue on "Green Grabbing." *Journal of Peasant Studies* 39(2).

Lappé, Francis Moore, Jennifer Clapp, Molly Anderson, Robin Broad, Ellen Messer, Thomas Pogge, and Timothy Wise. 2013. "How We Count Hunger Matters.' *Ethics & International Affairs* 27(3): 251–259.

Lavers, Tom. 2012. "'Land Grab' as Development Strategy? The Political Economy of Agricultural Investment in Ethiopia." *Journal of Peasant Studies* 39(1): 105–132.

Lisk, Franklyn, 2013. "'Land Grabbing' or Harnessing of Development Potential in Agriculture? East Asia's Land-based Investments in Africa." *Pacific Review* 26(5): 563–587.

Marchand, Philippe, Joel A. Carr, Jampel Dell'Angelo, Marianela Fader, Jessica A. Gephart, Matti Kummu, Nicholas Magliocca, Miina Porkka, Michael Puma, Zak Ratajczak, and Maria Cristina Rulli. 2016. "Reserves and Trade Jointly Determine Exposure to Food Supply Shocks." *Environmental Research Letters* 11(9): 095009.

McGahey, Daniel, Jonathan Davies, Nikolas Hagelberg, and Razingrim Ouedraogo. 2014. "Pastoralism and the Green Economy–a natural nexus," x, 58. Nairobi: IUCN and UNEP.

Mehta, Lyla, Gert Jan Veldwisch, and Jennifer Franco. 2012. "Introduction to the Special Issue: Water Grabbing? Focus on the (Re)Appropriation of Finite Water Resources." *Water Alternatives* 5(2): 193.

Mueller, Nathaniel, James Gerber, Matt Johnston, Ray Deepak, N. Ramankutty, and Jonathan Foley. 2012. "Closing Yield Gaps Through Nutrient and Water Management." *Nature* 490(7419): 254–257. DOI: 10.1038/nature11420.

Oberlack, Christoph, Laura Tejada, Peter Messerli, Stephan Rist, and Markus Giger. 2016. "Sustainable Livelihoods in the Global Land Rush? Archetypes of Livelihood Vulnerability and Sustainability Potentials." *Global Environmental Change* 41: 153–171.

Oki, Taikan, and Shinjiro Kanae. 2006. "Global Hydrological Cycles and World Water Resources." *Science* 313(5790): 1068–1072. DOI: 10.1126/science.1128845.

Raskin, Paul, Peter Gleick, Paul Kirshen, Gil Pontius, and Kenneth Strzepek. 1997. "*Water Futures: Assessment of Long-range Patterns and Prospects*." Stockholm: Stockholm Environment Institute.

Ravera, F., A. Scheidel, Jampal Dell'Angelo, G. Gamboa, T. Serrano, S. Mingorría, V. Cabello, N. Arizpe, and P. Ariza. 2014. "Pathways of Rural Change: An Integrated Assessment of Metabolic Patterns in Emerging Ruralities." *Environment, Development and Sustainability* 16(4): 811–820.

Ricciardi, Vincent, Navin Ramankutty, Zia Mehrabi, Larissa Jarvis, and Brenton Chookolingo. 2018. "How Much of the World's Food Do Smallholders Produce?" *Global Food Security* 17: 64–72.

Richter, Brian, 2014. *Chasing Water*. Washington, DC: Island Press.

Rosset, Peter. 2013. Food Sovereignty: Global Rallying Cry of Farmer Movements." *Food First Backgrounder* 9(4): 1.

Rulli, Maria Cristina, and Paolo D'Odorico. 2014. "Food Appropriation Through Large Scale Land Acquisitions." *Environmental Resources Letters* 9(6): 064030. DOI: 10.1088/1748-9326/9/6/064030.

Rulli, Maria Cristina, A. Saviori, and Paolo D'Odorico. 2013. "Global Land and Water Grabbing." *Proceedings of the National Academy of Sciences* 110 (3): 892–897.

Scheidel Arnim, Mario Giampietro, and Jesús Ramos-Martin. 2013. "Self-sufficiency or Surplus: Conflicting Local and National Rural Development Goals in Cambodia." *Land Use Policy* 34: 342–352.

Schellnhuber, Hans-Joachim, Arthur Block, Martin Cassel-Gintz, Jurgen Kropp, Gerhard Lammel, Wiebke Lass, and Gerhard Petschel-Held. 1997. "Syndromes of Global Change." *GAIA-Ecological Perspectives for Science and Society* 6(1): 18–33.

Seckler, David, Upali A. Amarasinghe, David James Molden, Radhika de Silva, and Randolph Barker. 1998. "World Water Demand and Supply, 1990 to 2025: Scenarios and Issues." *IWMI Research Report* 19. Colombo: IWMI.

Seekell, David, Joel A. Carr, Jampel Dell'Angelo, Paolo D'Odorico, Marianela Fader, Jessica Gephart, Matti Kummu, Nicholas Magliocca, Miina Porkka, Michael Puma, and Zak Ratajczak. 2017. "Resilience in the Global Food System." *Environmental Research Letters* 12(2): 025010.

Shiklomanov, Alexander I., 1991. "The World's Water Resources." Proceedings of the International Symposium to Commemorate 25 Years of the IHP, 93–126. Paris: UNESCO/IHP.

Skinner, Jamie, and Lorenzo Cotula. 2011. *Are Land Deals Driving 'Water Grabs.' Briefing: The Global Land Rush*. London: IIED. http://pubs.iied.org/1702IIED.html.

Smith, Michael D., Matthew P. Rabbitt, and Alisha Coleman-Jensen. 2017. "Who Are the World's Food Insecure? New Evidence from the Food and Agriculture Organization's Food Insecurity Experience Scale." *World Development* 93: 402–412.

Sosa, Milagros, and Margaret Zwarteveen. 2012. "Exploring the Politics of Water Grabbing: The Case of Large Mining Operations in the Peruvian Andes." *Water Alternatives* 5(2): 360.

Sullivan, Caroline A., Jeremy R, Meigh, and Anna Marie Giacomello. 2003. "The Water Poverty Index: Development and Application at the Community Scale." *Natural Resources Forum* 27(3): 189–199.

Temper, Leah. 2018. "From Boomerangs to Minefields and Catapults: Dynamics of Trans-Local Resistance to Land-Grabs." *Journal of Peasant Studies* 46(1): 1–29.

Third World Quarterly. 2013. Special Issue on Global Land Grabs. *Third World Quarterly* 34(9).

TLM. 2016. The Land Matrix Database. www.landmatrix.org/en/. Last Accessed May 2016.

UN. 1966. International Covenant on Economic Social and Cultural Right, Article 11, United Nations General Assembly December 16, 1966. Resolution 2200 A XXI.

Vorosmarty, Charles J., Pamela Green, Joseph Salisbury, and Richard B. Lammers. 2000. "Global Water Resources: Vulnerability from Climate Change and Population Growth." *Science* 289(5477): 284–288.

Von Braun, Joachim, and Ruth S. Meinzen-Dick. 2009. *"Land Grabbing" by Foreign Investors in Developing Countries: Risks and Opportunities*. Washington, DC: International Food Policy Research Institute.

Wagle, Subodh, Sachin Warghade, and Mandar Sathe. 2012. "Exploiting Policy Obscurity for Legalising Water Grabbing in the Era of Economic Reform: The Case of Maharashtra, India." *Water Alternatives* 5(2): 412.

Woodhouse, Philip, 2012. "Foreign Agricultural Land Acquisition and the Visibility of Water Resource Impacts in Sub-Saharan Africa." *Water Alternatives* 5(2): 208.

8 Food security in a premodern agrarian empire

The case of Rome

Kyle Harper

Introduction

For the satirist Juvenal, the phrase "bread and circuses" famously summed up the corrupt political culture of the Roman Empire (Juv., Satires, 10.81). In his eyes, the Roman *plebs*, or common folk, once upon a time full of republican virtue and hardiness, had traded freedom for comfort, duty for distraction, independence for handouts of grain. Juvenal was writing in the early second century, near the very height of the Roman Empire's power and prosperity, and his biting critique does point out the centrality of the food supply in the political economy of the empire. The population of the imperial capital had surpassed one million souls, the first city to reach such a peak, and the last Western city to do so until the nineteenth century. Providing the massive urban population with a reliable food supply was an imperative for any emperor who wished to remain in power for long, and it was a logistical feat that called upon the resources of a far-flung empire with an advanced administrative and legal system. The grain system of Rome was both a symbol of Rome's imperial power and a real artifact of the economic and political networks created by the empire (Rickman 1980; Erdkamp 2005).

In this chapter I will outline some of the main issues concerning the study of food security in the context of the Roman Empire in the first centuries of the Common Era. In fact, the provision of grain to the urban *plebs* is only the most-famous and best-understood link in a much-vaster network of agrarian production and distribution. The Roman Empire was home to an aggregation of thoroughly premodern/preindustrial societies. Most people were peasants, most people lived in the provinces, and most people were poor. Across the distance of millennia, we know very little about their experience of life. At the same time, the Roman Empire is one of the ancient polities best known to us. It affords us a chance to reconstruct, often obliquely and imperfectly, something of the conditions of subsistence during an important phase of the human past, in the context of highly complex but still agrarian societies. Usually we see the Roman Empire through the prism of an elite perspective, but even still, issues of food scarcity and supply impinged on the aristocracy's view of the world. And as the words of a humble prayer from a far corner of the empire remind us—"give us this day our daily bread"—the food security concerns of the ordinary inhabitants of the empire are not totally unknown to us. At the end of the period we will consider in this chapter, the rise of Christianity started to alter the politics of hunger, dignity, and entitlement in lasting ways.

The dimensions of the Roman Empire

The Roman Empire was an extraordinary political formation in both its temporal and geographical scope. The empire was acquired during the last centuries before the Common Era, under the political regime of the Roman republic, the *res publica* (Mouritsen 2013). In this period, political authority was held not by a monarch but through a dynamic constitutional system that distributed power between the senatorial aristocracy and the common citizenry (the *plebs*). The republican constitution facilitated the mobilization of military power on a massive scale, which fueled the expansionary conquests that led to the acquisition of new territory. The growth of the state in turn catalyzed social change and destabilized the constitutional regime; elite competition for power triggered a century of intermittent civil war leading to the new political regime imposed by Julius Caesar and his adopted son, the future emperor Augustus. In the words of the classicist Mary Beard, "The empire created the emperors—not the other way round" (Beard 2015). Augustus (r. 27 BCE–14 CE) established the imperial regime in which a monarch in all but name effectively monopolized control of the state and held supremacy over the senatorial class and the Roman legions alike. This regime endured in its basic form for more than four centuries in the western territories, even longer in the eastern realms. Measured by sheer longevity, it was one of the most successful political regimes in human history.

The spatial scale of the empire was even more exceptional (Scheidel 2014). The northern tips of the empire reached across the 56th parallel; the southern territories of the empire stretched below 24° latitude. The western boundary of the empire was the impassable Atlantic Ocean, while in the east the empire controlled the humid coastal Levant and faded out in the arid regions beyond. This huge formation encompassed an astonishing variety of ecological regimes (Harper 2017). At the core of the empire lay the Mediterranean Sea and its associated ecologies, defined by cool wet winters and hyper-arid summers. Much of the west was oceanic and continental, whereas desert defined the southern and eastern frontiers of the empire along the subtropical ridge. Egypt, furthermore, was a special case, due to the unique ecology of the Nile valley. The annual inundation, fed by rainwaters from highland Ethiopia, made the ribbon of cultivated land cutting through the desert miraculously fertile. Egypt, as we will see, was the breadbasket of the empire. The ecological diversity and complexity of the Roman Empire frames the dialectics of food security and insecurity.

Food and society in the Roman Empire

In the heartlands of the empire, the "Mediterranean triad" of wheat, wine, and oil dominated the diet, complemented by some meat (with a Roman predilection for pork), as well as fish, legumes, and garden and tree crops. For most inhabitants of the empire, cereals must have been the main source of calories on a daily basis, although there has been debate about the extent to which the ordinary citizen of the Roman Empire was undernourished due to an overreliance on a monotonous diet consisting predominantly of cereals (Garnsey 1988). The question is complex, and the answer must have varied over time, space, and social class. In general, comparative evidence assures us that most Romans lived relatively close to the subsistence level, at per capita income levels equivalent to something like $800 to $1,000 per year or less, in modern terms (i.e., in 1990 Geary-Khamis dollars, see Scheidel and Friesen 2009). According to one careful analysis, we could broadly imagine Roman society in three tiers of wealth: a wealthy aristocracy making up approximately 2 percent of the population; a "middling" element, living comfortably above subsistence that accounted for approximately10 percent of the population; and "the rest"—the numerically dominant for whom basic subsistence was a constant concern (Scheidel and Friesen 2009).

In the Roman world there was a preference for durum wheat, *triticum durum*, so much so that access to it was considered something of a privilege (Erdkamp 2013). A little unexpectedly, one of our greatest sources for the interplay of diet and social class in the Roman Empire is the famous Greek doctor, Galen, who flourished in the later second and early third centuries. Galen came from a modestly elite family in Pergamum, a Greek town in Asia Minor that, like so many, thrived in the Roman Empire (Schlange-Schöningen 2003; Mattern 2013). Galen was a keen observer of social life, and he explored the empire in search of cures and remedies. He was also a prolific writer. He describes for us the rough black bread eaten by the peasants who lived just beyond the orbit of a city like Pergamum. Barley, einkorn, emmer, and other grains were hardy and reliable, if less appetizing to an urbanite like Galen, whose digestive system was upset by the rough bread of the peasantry. Galen gives us a rare but vivid sense of the radically distinct foodways that could exist in close physical proximity inside the Roman Empire (Garnsey 1988; Sallares 1991; Harper 2017: 56).

Traditionally, the study of ancient diet and nutrition has relied on fragmentary literary or documentary sources, but increasingly, bioarchaeological evidence is filling in the picture. There are probably hundreds of thousands of Roman-period skeletons in museum collections and new discoveries practically every month. Bones—their size, shape, scars, and chemistry—are evidence of how people lived and what they ate (Killgrove 2014). Stable isotope analysis is particularly important for the study of past diets (Prowse 2004; Killgrove 2010) because stable isotopes of carbon and nitrogen cycle through nature before being processed by the body into bone collagen. The ratio of heavy isotopes to normal atoms preserved in bones can reflect the composition of an individual's diet. For instance, some plants, like millet, use the C_4 carbon fixation pathway, which uses proportionately more heavy carbon isotopes, and a Roman skeleton with enrichment for heavy carbon isotopes might reflect a hardy peasant diet. Enrichment with heavy nitrogen is a signal that dietary protein was consumed from organisms higher up the food chain. Perhaps one of the early signals from Roman bioarchaeology is that many individuals have nitrogen isotope ratios suggesting moderate levels of terrestrial animal sources of protein and marine sources of protein (Prowse 2004; Craig et al. 2009; Cummings 2009; Rutgers et al. 2009; Killgrove 2010). Roman Britain has been most heavily studied, and some meat and fish were part of the diet. Teeth, too, can reflect diet. Dental caries is a pathology caused by a narrow diet of cereal-derived carbohydrates, and some though not all Roman populations studied had high levels of this pathology (Manzi et al. 1999; Bonfiglioli, Brasili, and Belcastro 2003; Cucina et al. 2006).

At present, the bioarchaeological record of the Roman Empire is more promising than conclusive; much more work needs to be done to draw firm conclusions about the complexity of Roman health and diet and the patterns of change over time. But if there is one interesting and repeatedly validated signal coming from Roman bones, it is that the Romans were short (Harper 2017: 75–80, for an overview). Both at the center of the empire, and in the periphery (in this case again, in Britain), the Romans were shorter than their Iron Age predecessors and their medieval successors (Giannecchini and Moggi-Cecchi 2008, with Harper 2017: 75–77 on why this study is so valuable; Figure 8.1).

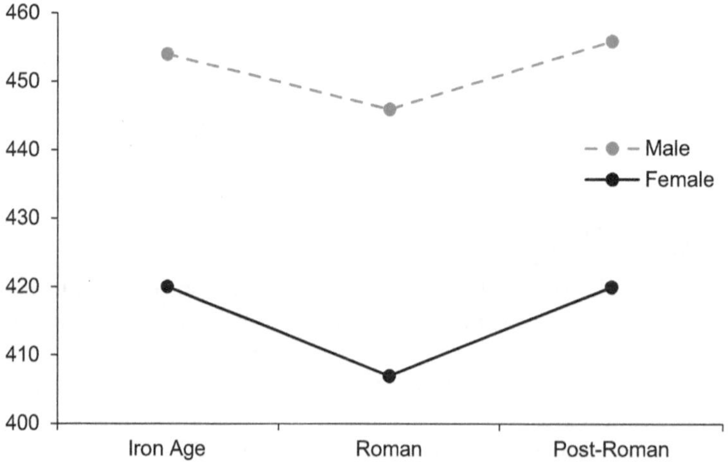

Figure 8.1 Average femur length in Italy (mm).
Source: Giannecchini and Moggi-Cecchi 2008.

Achieved physical stature is a result both of genetics and of environment; bone length is thus a crude proxy of biological well-being during an individual's developmental period (lasting more or less until the age of twenty). The relevant environmental factors include diet, disease, and stress, which of course are highly confounding (e.g., nutritional deprivation leads to greater disease susceptibility). It is an open question why average stature declined in the Roman period. I have argued that an intensifying disease burden was the primary factor. The argument falls into two parts. First, the Romans built a highly urbanized and interconnected empire that was conducive to the transmission dynamics of infectious pathogens. There are solid comparative examples, particularly from the early stages of industrialization, where economic development worsened biological well-being, largely because health was adversely affected by early urbanization. Second, I argue that Roman economic development outpaced demographic growth; in other words, the Roman Empire did *not* experience Malthusian impoverishment, so we cannot look to a population that had overrun the ecological carrying capacity of its resource base as the explanation for the unhealthy Romans.

The strictly Malthusian view, which I reject, certainly has a baseline of plausibility, and the issue is highly relevant for the broader question of food security (see Erdkamp 2016 for a recent discussion). There is no doubt that the first two centuries of the Roman Empire were a phase of strong population growth, even if absolute figures remain uncertain (Harper 2017: 29–33). At the beginning of the reign of Augustus, there may have been something like 60 million people living in the territory that was or would become Roman, out of a global *Homo sapiens* population of 250 million; at its peak under Marcus Aurelius, the Roman population was more like 75 million (out of a global 300 million, so that one in four people may have lived inside the empire). This rate of growth is highly credible. In fact, in the two centuries leading up to the Black Death, with similar levels of technology and lower levels of peace, European populations grew even faster, roughly doubling over the course of two centuries (Campbell 2016). In the medieval case, demographic expansion put pressure on agrarian resources and led to widespread food crises and declining real wages, which were only reversed by the catastrophic population losses in the bubonic plague pandemic. So, did the Roman Empire experience similar Malthusian pressures and, if not, how were they avoided against a background of robust demographic expansion?

Although archaeologists are generally impressed by the vestiges of Roman economic performance (put simply, the Romans made, consumed, and left behind a lot of stuff), definitive proof that per capita incomes were stable or rising is limited (Scheidel 2009). The archaeological record could, after all, merely reflect the greater output of a much larger, but on average poorer, population. But there are important indications of genuine economic development in the early empire. The urbanization rate may have been as high as 20 percent, which is quite exceptional for a large, premodern polity, and could only have been supported by high levels of specialization and exchange (Wilson 2011; Hanson 2016). Moreover, there is evidence for rising real wages in Roman Egypt, where the arid climate has preserved an abundant papyri record. These data show that the wages of unskilled laborers rose faster than the price of foodstuffs, suggesting that even ordinary inhabitants of the Roman Empire were beneficiaries of a more developed economy (Harper 2016a). The evidence from Egypt contradicts the Malthusian hypothesis.

The Romans were able to achieve intensive growth through the classical mechanisms of technology and trade, through probably mostly the latter (Temin 2013). The Roman agrarian economy was based on the basic Iron Age technology package: draught animals for traction, wheels, and iron plows. Larger and more sophisticated milling machinery probably enhanced productivity; watermills, for instance, were more widespread than once believed (Wilson 2002). Overall, Roman technical advances were far from revolutionary, but capital investment and technical diffusion (rather than innovation) may have considerably enhanced productivity in the primary economic sector. Roman economic gains were probably more driven by the expansion of trade (Harper 2017: 36–38). The empire created a massive and safe "free-trade" zone. There were advances in transport technology (bigger and safer ships) and rather famous civil engineering (roads, obviously, but also harbors). Advanced institutions of credit as well as Roman law, with strong protections of private property (the nearly absolute concept of *dominium*, for instance, and the wide discretion of testators) facilitated exchange. In short, lower transaction costs, favorable institutions, and ease of transportation all fostered economic exchange and integration. Consequently, the Roman economy grew, even on a per capita basis, in the long phase of growth between Augustus and Marcus Aurelius (r. 161–180), when a period of crisis began.

Economic development, interregional integration, and public institutions were all important in the food-supply network of the Roman Empire. In some ways, certainly in scale, the city of Rome was an unusual case. It was the metropole of the empire, and its citizens were the beneficiaries of imperial rents and entitlements that went back to the politics of the *plebs* in the late republic. In 123 BCE, the populist leader Gaius Gracchus carried through reforms that controlled the price of grain on the market in Rome, guaranteeing its affordability. An entitlement was established that would last more than half a millennium. In some form or other, access to food became a privilege of the people of Rome. In its classic form (pilloried by Juvenal), the grain dole was a monthly distribution of unprocessed grain to Rome's citizens. It was a primary responsibility of the emperors, and though dwarfed by the cost of provisioning the army, it was a major item of public expense. After defeating his rivals Mark Antony and Cleopatra, Octavian/Augustus seized Egypt, and the rich province was considered the personal possession of the emperors. The grain taxes collected in Egypt were a principal source of the food for Rome. Each year, a huge flotilla of deep-hulled grain ships sailed from Alexandria to Rome. There has been extensive debate about the relative role of market transfers versus state transfers in the grain supply of Rome (Erdkamp 2013). Certainly, there was an extensive private trade in grain, too, and the Roman state seems to have subsidized or incentivized private merchants to carry grain to Rome, where it was purchased by the state for distribution. Over time, the system evolved. From the third century, bread rather than grain was distributed, and then wine and pork were added to the public supply system or guaranteed at low rates (Sirks 1991). In the fourth century, the state relied less on the open market for purchase, and more on provision and transport by command. By this time, real power had drifted to other capitals nearer the frontiers, but the symbolic power of Rome meant that its *plebs* retained outsized political entitlements. Unsurprisingly, Constantine established parallel privileges in "New Rome" (i.e., Constantinople).

The city of Rome could not have sustained a population of one million people without extremely robust systems of production, transport, storage, and distribution, underwritten by a powerful mix of private market forces and state supervision. Mention should be made of the stunning grain-storage facilities in the largest Roman cities. It was a mark of his frugality and sound management that when the emperor Septimius Severus died, he had enough grain stored to feed the people for seven years (Hist. Aug. Vita Sev. 23.6). Rome was unique in its scale and the particular configuration of political entitlements, but other cities had their own institutions. We know most about the Greek cities under Roman rule (thanks to literacy rates and habits of inscribing texts on stone). Greek towns had public officials in charge of the grain market, and some had public distributions of grain to citizens (Zuiderhoek 2008). Maybe most importantly, there was a deeply engrained cultural tradition of *euergetism* (simply defined, "doing good deeds"). Civic elites were strongly expected to be public benefactors (Zuiderhoek 2009). The wealthy were expected to practice munificence for the benefit of their city, for instance in the form of building monumental structures like theaters and baths. But providing food at no cost or at artificially cheap rates to fellow citizens, especially during times of shortage, was also a major part of the implicit social compact between the civic elites and the mass of ordinary citizens.

Increasingly, the Roman emperors became more visible as the patrons-in-chief, lavishing generosity on the towns under their rule, particularly in times of crisis. The emperor Trajan would

> divert and direct the abundance of the land now here, now there, as the moment and necessity demand. He would feed and protect a rescued nation across the sea as though it were some part of the Roman people and plebs.
>
> (Pliny, *Pan.* 32)

> Hadrian had seen many cities, more than any other emperor, and he 'took care of them all,' so to speak, giving water to some, harbors to others, grain to some, public works to others, money to some, and honors to others.
>
> (Cassius Dio, *Hist. Rom.* 69.5.3)

In sum, the food-supply system of the high Roman Empire was framed by the interplay of the free market and cultural institutions oriented around ideas of civic membership and personal status. Access to public provision was based on civic identity, legal status, and systems of patronage.

Climate, resilience, and crisis

The climate of the empire's core region required forms of social resilience (Garnsey 1988; Sallares 1991; Horden and Purcell 2000). The Mediterranean climate is predictably unpredictable. The semi-arid ecology depends on the rainy season for most of its precipitation. Many areas that were inherently close to the minimal threshold for agriculture were cultivated in the Roman period, and population growth only encouraged the exploitation of ever more marginal land. Perhaps every four or five years, the rain was insufficient for the wheat harvest, and food became scarce on at least the local level (Garnsey 1988). Most ancient Mediterranean peasants would have lived through times of food shortage. Consequently, strategies evolved to manage risk. Various grains, including hardier varietals, were always planted. Food was dried, pickled, and stockpiled by any possible means. Emergency crops could be sown quickly in times of crisis. Social institutions like clientage or informal lending were important in peasant economies. Although we do not have much information about subsistence migration in the Roman world, we do know that cities, particularly Rome itself, were magnets for immigrants and not just adult males. We need more and better models in the study of the ancient world to help us understand how resilient farmers might be against the inevitability of interannual climate variability, but it is undeniable that highly resilient lifeways were simply a condition of existence in the premodern Mediterranean.

The politics of food insecurity were probably always in the background in a Mediterranean society that depended both on the fickle climate and on complex networks of distribution and exchange. Gregory Aldrete counted 154 episodes of riot in the city of Rome between the late republic and late empire (200 BCE–375 CE), with rising food prices the leading cause of political disturbance (Aldrete 2013). Similarly, in the provinces, the cities seem to have been the focal point of the politics of food security (Stathakopoulos 2004). It is telling that migration to the city was a common strategy even for desperate peasants in times of harvest failure: as the centers of political administration, the cities were sites of food storage, emergency distribution, and relief.

The most striking conclusion of Peter Garnsey's classic *Famine and Food Supply in the Graeco-Roman World* is that true famine was exceedingly rare in the Roman Empire. Food shortages and times of scarcity were commonplace, but famine or food crisis inducing mass deprivation was anything but common. Certainly, in the high Roman Empire, there are few indications of large-scale food crises. If the absence of evidence is telling (and it is always possible that events affecting the poor and marginal are underrepresented), what accounts for the relative stability of the food-supply systems of the Roman period? The answers may be both environmental and social-institutional.

On the environmental side, the ever-growing body of paleoclimate evidence is deepening our understanding of the history of climate change and variability. The last approximately twelve thousand years, known as the Holocene Epoch, have been a time of warmth and stability relative to the wild oscillations of the preceding Pleistocene. But even in the Holocene, the climate has been changing at various scales—millennial, centennial, decadal, and so on. Paleoclimate proxies—tree rings, ice cores, cave minerals, marine and lake sediments, pollen records, and so on—can let us explore the climate's past when calibrations between physical records and climate parameters can be established. Sometimes, proxies provide a record at extremely high temporal resolution—annual or even subannual. Our picture of the Roman climate is starting to become clearer.

In the climate literature, the last centuries of the Roman republic and the first centuries of the imperial period (ca. 200 BCE–150 CE) are known as the Roman Climate Optimum (Harper 2017; Harper and McCormick 2018). Although it is misleading to imagine that the climate can be "good" or "bad" across large stretches of time and space, this period of climate history seems in general to have been favorable to agricultural production in much of the Mediterranean region. It was a highly stable period, with elevated levels of solar radiation and only one major volcanic event (in 44 BCE, the year of Julius Caesar's assassination). Temperatures were warmer but humidity seems to have been high and stable relative to the modern period, which would have been a boon to Roman farmers, although the hydroclimate is more regionally variable, and the evidence at present is much weaker. And it also seems that the Nile flood was regular and dependable in this period. The overall stability of the climate in these centuries may have been an enabling factor in Roman growth and acted to suppress some of the worst extremes of interannual variability that can induce food crisis. The period that followed, by contrast, was marked by more extreme variability and by a cooler, more arid climate in central regions of the empire. And the Nile seems to have become more erratic, too, occasionally depriving the entire empire of its great insurance policy in food production. A major failure of the Nile in the 240s clearly induced an acute food shortage and might have helped to catalyze a broader period of crisis across the empire (Harper 2017: 132–134).

But the absence of acute and large-scale food crises in the first centuries of the empire cannot be explained by environmental variables alone. Social and institutional factors worked in concert with the natural environment to stabilize the food system. The interregional integration of the empire may have been the most important factor of all. According to a fourth-century source, bad harvests could be easily overcome because "provinces come to the relief of each other, good harvests here supplying the deficiencies of bad harvests there" (Symmachus, *Rel.* 3.17). Moreover, strong imperial and municipal institutions buffered against acute crisis, at least for the politically important and visible urban populations. Roman governors intervened against price gouging. Flourishing civic elites could be relied upon in times of scarcity.

Signs of stress started to appear in the later second century. A complex environmental, biological, and political crisis unfolded in the reign of Marcus Aurelius. The stability of the Roman Climate Optimum came to an abrupt end. A pandemic disease event known as the Antonine Plague induced mortality on an unprecedented scale; possibly the debut of the smallpox virus, the pandemic spread over the entire empire and carried off millions of victims. Marcus survived an attempted usurpation, but barbarians broke through the northern frontier and pierced deep into the soft tissue of the empire—all a harbinger of a challenged future. And, finally, we have compelling evidence for systemic failure in the food system, again from Galen, who lived through these events. Galen reported "continuous famine for not a few years among many of the nations subject to Rome." Hungry town-dwellers descended on the countryside, and "in accordance with their universal practice of collecting a sufficient supply of wheat to last a whole year," stripped the fields, leaving the rustics to scavenge and survive on twigs and grass (Galen, *Bon. Mal. Succ.* 1, with Harper 2017: 330).

The empire came through the crisis under Marcus intact, but the episode underscores the complex relationships between the physical climate, the biological environment, and food production. In a society where infectious diseases were always the leading cause of death, the physical environment was a strong control on the incidence of morbidity and mortality. Through many different mechanisms, the climate was a major factor in the biological environment (Harper 2018). One pathway of influence was through the system of food production: drought might induce harvest failure, depressing the nutritional status of a population; malnutrition weakens the metabolically expensive immune system and increases susceptibility to infectious disease; elevated morbidity and mortality can undermine the availability of labor and further depress agricultural production. In short, the food system is intricately embedded in nonlinear systems that include the physical and biological climates.

We can follow these complex chains of causality in an unusually well-attested event from the eastern fringes of the Roman Empire in the late fifth century. The small-scale but acute crisis occurred in the town of Edessa and was vividly described by an eyewitness (see Harper 2018). In 494–95 CE, a plague befell "all" the people of the city.

> Swellings and tumours appeared on all our citizens, and the faces of many became puffed up and filled with pus, making a fearful sight. Some had sores or pustules over their whole body, even to the palms of their hands and the soles of their feet
>
> (Pseudo-Joshua, *Chron.* 26, 38–44: see Trombley and Watt 2000:23)

(The symptoms strongly suggest a smallpox epidemic.) In March 500 CE, there was a plague of locusts. In April, grain scarcity set in, and prices rose. In the early summer, millet (an emergency crop) was sown, but it failed. True famine arrived. Migrants came in droves. To fight starvation, the poor ate anything they could find. The hunger caused mortality to worsen as it turned to winter. The public authorities were overwhelmed. The death toll mounted as a pestilence flared across the region. Now, we are told, even the rich died. It is a rare description of a systemic failure of a resilient system, overwhelmed by a combination of environmental and biological shocks.

Food crises probably were in fact more frequent and more intense in the later centuries of the Roman Empire (Stathakopoulos 2004). The physical climate was more variable. The disease environment was more unhealthy and prone to large mortality events. Imperial and municipal institutions may have been weaker or more rigid than before. But it is also true that our *evidence* for food insecurity is richer in this period because the cultural frame had shifted with the rise of Christianity. We know more about food crisis, because the Christian church was concerned with issues of poverty and hunger in ways that were frankly novel.

The conversion of Constantine in 312 CE accelerated the spread of the religion and set it on a path to social hegemony. Within a generation, we have evidence from bishops that provide a level of insight and detail that was lacking before. The classic example comes from a food shortage that occurred in Caesarea, in Cappadocia, during the tenure of the famous church father Basil (Holman 2001). He preached a series of sermons as the region steeled itself for scarcity. For Basil, the food shortage revealed the sharp injustice of the social order. In a pathetic scene that called forth all of Basil's rhetorical training, he conjured the image of a poor father, forced to decide which of his children to sell in order to stave off starvation.

> How can I put before your eyes the suffering of the poor? … He turns his eyes at last to his children, to take them to the market and find a way to put off death. … Imagine his deliberations. "Which will I sell first? Which will the grain merchant like best?" … He goes, and with ten thousand tears, sells his most beloved son.
>
> (Basil of Caesarea, *Dest. Horr.*, 4)

Basil also acted to relieve the crisis through his own ecclesiastical programs.

As Peter Brown has shown, the Christian leaders of the fourth century effectively invented "the poor" as a visible social category (Brown 2002). The social order imagined by the bishops cut across some very old ideas, and in unexpected ways. No longer were entitlements framed by civic identity and status, but by an ideal of common humanity. This imaginary social order implied new models of leadership and new models of human solidarity. Basil's brother, the bishop Gregory of Nyssa, preached eloquently about the hungry who were suffering right in the midst of the city. He insisted that, as bearers of the image of God, they carried the same dignity as the rich (Harper 2016b). Gregory's sermons ask his audience, literally, to look into the dark corners of the city and take pity on the sick and starving. Christian charity was replacing civic entitlement as a defining social ideal, and this new ideal would endure the failure and transformation of Roman political institutions in the passage to the Middle Ages.

Conclusion

Did greater food instability contribute to the decline of the Roman Empire? The evidence does not allow firm claims, and the relationship between agrarian production and imperial power is immensely complex, but it is not unreasonable to consider the possibility. The paleoclimate evidence increasingly confirms that the period ca. 530–680 CE should be known as the "Late Antique Little Ice Age," a phase of generally colder weather across much of the Northern Hemisphere (Büntgen 2016). Across much of Europe and the Mediterranean, this climate would have been less favorable to agricultural output and stability, weakening the economic core of the empire. Farmers no doubt adapted, but in aggregate the climate is a powerful factor in agricultural output. The question continues to be explored, and better paleoclimate evidence will continue to enrich our knowledge of this distant period. Moreover, the role of food insecurity *beyond* Rome's borders, including on the Eurasian steppe, in pushing mass migration during late antiquity, is a major and intriguing possibility that continues to be debated (Cook 2013).

What can be said with certainty is that the collapse of the Roman Empire meant the end of the ancient food distribution systems on which the political culture of "bread and circuses" had been based for hundreds of years. In the fifth century, the food-supply system of ancient Rome broke down, and the population of the city started to dwindle. The food shipments from Egypt to Constantinople only ended in the seventh centuries, in the wars with Persia and then the Arab conquests. Symbolically, and to a certain extent materially, the end of these trans-Mediterranean systems of urban food security marked the end of an ancient order.

References

Aldrete, Gregory. 2013. "Riots." In Paul Erdkamp, ed. *The Cambridge Companion to Ancient Rome*. 425–440. Cambridge: Cambridge University Press.

Beard, Mary. 2015. *SPQR: A History of Ancient Rome*. New York: Norton.

Bonfiglioli, B., P. Brasili, and M. G. Belcastro. 2003. "Dento-alveolar lesions and nutritional habits of a Roman Imperial Age population (1st–4th c. AD): Quadrella (Molise, Italy)." *Homo* 54(1): 36–56.

Brown, Peter R. L. 2002. *Poverty and Leadership in the Later Roman Empire*. Hanover, NH: Brandeis University Press.

Büntgen, Ulf et al. 2016. "Cooling and societal change during the Late Antique Little Ice Age from 536 to around 660 AD." *Nature Geoscience* 9: 231–236.

Campbell, Bruce. 2016. *The Great Transition: Climate, Disease and Society in the Late-Medieval World*. Cambridge: Cambridge University Press.

Cook, Edward. 2013. "Megadroughts, ENSO, and the invasion of Late-Roman Europe by the Huns and Avars." In William V. Harris, ed. *The Ancient Mediterranean Environment between Science and History*, 89–102. Leiden: Brill.

Craig, Oliver E. et al. 2009. "Stable isotopic evidence for diet at the Imperial Roman coastal site of Velia (1st and 2nd Centuries AD) in Southern Italy." *American Journal of Physical Anthropology* 139(4): 572–583.

Cucina, Andrea et al. 2006. "The necropolis of Vallerano (Rome, 2nd–3rd Century AD): An anthropological perspective on the Ancient Romans in the *Suburbium*." *International Journal of Osteoarchaeology* 16(2): 104–117.

Cummings, Colleen. 2009. "Meat consumption in Roman Britain: The evidence from stable isotopes." In Driessen, Mark et al., eds. *TRAC 2008: Proceedings of the Thirteenth Eighteenth Annual Theoretical Roman Archaeology Conference, Amsterdam 2008*, 73–83. Oxford: Oxford University Press.

Erdkamp, Paul. 2005. *The Grain Market in the Roman Empire: A Social, Political and Economic Study*. Cambridge: Cambridge University Press.

Erdkamp, Paul. 2013. "The Food Supply of the Capital." In P. Erdkamp, ed. *The Cambridge Companion to Ancient Rome*, 262–277. Cambridge: Cambridge University Press.

Erdkamp, Paul. 2016. "Economic growth in the Roman Mediterranean world: An early good-bye to Malthus?" *Explorations in Economic History* 60: 1–20.

Garnsey, Peter. 1988. *Famine and Food Supply in the Graeco-Roman World: Responses to Risk and Crisis*. Cambridge: Cambridge University Press.

Giannecchini, M., and J. Moggi-Cecchi. 2008. "Stature in archeological samples from Central Italy: Methodological issues and diachronic changes." *American Journal of Physical Anthropology* 135(3): 284–292.

Hanson, J. W. 2016. *An Urban Geography of the Roman World, 100 BC to AD 300*. Oxford: Oxford University Press.

Harper, Kyle. 2016a. "People, plagues, and prices in the Roman world: The evidence from Egypt." *Journal of Economic History* 76(3): 803–839.

Harper, Kyle. 2016b. "Christianity and the roots of human dignity in late antiquity." In T. Shah and A. Hertzke, eds. *Christianity and Freedom, Volume I: Historical Perspectives*, 123–148. Cambridge: Cambridge University Press.

Harper, Kyle. 2017. *The Fate of Rome: Climate, Disease, and the End of an Empire*. Princeton, NJ: Princeton University Press.

Harper, Kyle. 2018. "Invisible environmental history: Infectious disease in late antiquity." In Adam Izdebski and Michael Mulryan, eds. *Environment and Society in the Long Late Antiquity*, 118–134. Leiden: Brill.

Harper, Kyle, and Michael McCormick. 2018. "Reconstructing the Roman Climate." In W. Scheidel, ed. *The Science of Roman History*, 11–52. Princeton, NJ: Princeton University Press.

Holman, Susan. R. 2001. *The Hungry Are Dying: Beggars and Bishops in Roman Cappadocia*. Oxford: Oxford University Press.

Horden, Peregrine, and Nicholas Purcell. 2000. *The Corrupting Sea: A Study of Mediterranean History*. Oxford: Oxford University Press.

Killgrove, Kristina. 2010. *Migration and Mobility in Imperial Rome*, PhD diss., University of North Carolina.

Killgrove, Kristina. 2014. "Bioarchaeology in the Roman Empire." In C. Smith, ed., *Encyclopedia of Global Archaeology*, 876–882. New York: Springer.

Manzi, Giorgio, et al. 1999. "Discontinuity of life conditions at the transition from the Roman Imperial Age to the Early Middle Ages: Example from Central Italy evaluated by pathological dento-alveolar lesions." *American Journal of Human Biology* 11(3): 327–341.

Mattern, Susan. P. 2013. *The Prince of Medicine: Galen in the Roman Empire*. Oxford: Oxford University Press.

Mouritsen, Henrik. 2013. "The Roman Empire I: The Republic." In P. F. Bang and W. Scheidel, eds. *Oxford Handbook of the State in the Ancient Near East and Mediterranean*, 383–411. Oxford: Oxford University Press.

Prowse, Tracy. 2004. "Isotopic paleodiet studies of skeletons from the Imperial Roman-age cemetery of Isola Sacra, Rome, Italy." *Journal of Archaeological Science* 31(3): 259–272.

Rickman, Geoffrey. 1980. *The Corn Supply of Ancient Rome*. Oxford: Oxford University Press.

Rutgers, Leonard. V. et al. 2009. "Stable isotope data from the early Christian catacombs of Ancient Rome: New insights into the dietary habits of Rome's early Christians." *Journal of Archaeological Science* 36(5): 1127–1134.

Sallares, Richard. 1991. *The Ecology of the Ancient Greek World*. Ithaca, NY: Cornell University Press.

Scheidel, Walter. 2009. "In search of Roman economic growth." *Journal of Roman Archaeology* 22: 46–70.

Scheidel, Walter. 2014. "The shape of the Roman world: Modelling imperial connectivity." *Journal of Roman Archaeology* 27: 7–32.

Scheidel, Walter, and J. Steven Friesen. 2009. "The size of the economy and the distribution of income in the Roman Empire." *Journal of Roman Studies* 99: 61–91.

Schlange-Schöningen, Heinrich. 2003. *Die römische Gesellschaft bei Galen: Biographie und Sozialgeschichte*. Berlin: De Gruyter.

Sirks, Boudewign. 1991. *Food for Rome: The Legal Structure of the Transportation and Processing of Supplies for the Imperial Distributions in Rome and Constantinople*. Amsterdam: J. C. Gieben.

Stathakopoulos, Dionysios. 2004. *Famine and Pestilence in the Late Roman and Early Byzantine Empire: A Systematic Survey of Subsistence Crises and Epidemics*. Burlington, VT: Ashgate.

Temin, Peter. 2013. *The Roman Market Economy*. Princeton, NJ: Princeton University Press.

Trombley, Frank and John Watt, eds. 2000. *The Chronicle of Pseudo-Joshua the Stylite*. Liverpool: Liverpool University Press.

Wilson, Andrew I. 2002. "Machines, power and the ancient economy." *Journal of Roman Studies* 92: 1–32.

Wilson, Andrew. 2011. "City sizes and urbanization in the Roman Empire." In A. Bowman and A. Wilson, eds. *Settlement, Urbanization, and Population*, 161–195. Oxford: Oxford University Press.

Zuiderhoek, Arjan. 2008. "Feeding the citizens: Municipal grain funds and civic benefactors in the Roman East." In R. Alston and O. van Nijf, eds. *Feeding the Ancient Greek City*, 159–180. Leuven: Peters.

Zuiderhoek, Arjan. 2009. *The Politics of Munificence in the Roman Empire: Citizens, Elites and Benefactors in Asia Minor*. Cambridge: Cambridge University Press.

9 The transformation of famine relief regimes in modern China

Pierre Fuller

Introduction

China has shed a longstanding reputation as a "land of famine" (Mallory 1926) thanks to large gains in wealth and food production in the past few decades. Meanwhile, Chinese institutions have become major players in the aid and development sector in parts of the world, such as Africa, troubled more recently by war, severe poverty, and food scarcity. Scholars have underlined and criticized the significant role of the Chinese state in aid operations generally, both domestically and internationally (Brautigam 2009; Hirono 2013). But is the state integral to Chinese humanitarian action? What was its role in the past? Who assumed the burden of relief provisioning? What principles lay behind humanitarian action?

This chapter, using the example of famine, contributes to mounting interest in the humanitarian legacies of the Global South by casting light on the diverse range of actors who historically have occupied the humanitarian space in China. In the process, it touches on fundamental issues of state power, relief agency, and humanitarian principles that are relevant to contemporary Chinese aid ventures and will undoubtedly remain so in the years ahead.

This chapter examines four famines in modern China, defined as the period since the First Opium War of 1839–42. First it assesses the disastrous famine of 1876–79 and the successful famine relief effort of 1920–21 to highlight the following factors:

- the internationalisation of relief in the late nineteenth and early twentieth centuries;
- the multilayered nature of China's relief system, incorporating official and nonofficial efforts at all levels; and
- the persistence of older, indigenous relief practices in China well into the twentieth century.

Second, it shows how, during the famine of 1928–30, civil war and the changing priorities of the Chinese state contributed to the collapse of a relief regime that had worked relatively well earlier in the decade. Third, it analyses the Great Leap Forward famine of 1958–62, which produced the highest death toll of any such event in modern history, and which marked a culmination of radical changes in the priorities of the modern Chinese state and society.

From the discussions of these famines two key points emerge. First, the recent surge in Western and Chinese attention to the Great Leap famine is hugely important (Thaxton 2008; Dikotter 2011; Yang 2012) but it risks reducing public perceptions of the long and varied Chinese experience of famine to its mid-twentieth-century Maoist form. (Of the three other famines covered here, each of which was akin to a full-scale war in terms of human destruction, only one has a book devoted to it in English until recently.) In addition, the *pre*-Maoist model of relief in China was not strictly statist, and earlier relief regimes are becoming relevant again as Chinese society slowly reacquires the elements of a civil society it possessed before the Communist Revolution of 1949. This development has prominently played out in the broad-based response to the Sichuan earthquake of 2008 and nationwide droughts of 2010. Second, Western (and in some cases Chinese) accounts of past disaster relief in China have overemphasized international, largely Western, engagement at the expense of indigenous relief efforts. Institutions and relief groups operating over long distances generate far more of the publicity and record keeping later used by historians than do informal or temporary relief agents originating in or near the disaster field. This is not unique to China, and arguably is a universal problem that can lead to an understanding of disaster events that exaggerates the roles of international or exogenous relief agencies. Hence, this chapter looks beyond central state policy to include various examples of local and regional relief that continue to be employed to this day.

In the space available it is impossible to address a host of critical issues related to disaster relief, including corruption, disease control, poverty, profiteering, and refugee movement, or the principles behind humanitarian action in any detail. Rather, the aim is to produce a sketch of the changing relief regimes that kept people from dying, or failed to do so, using the example of drought in modern China.

Lastly, it is important to include a cautionary word on numbers, which rarely fail to be massive in the case of China. Regularly rounded by the million, mortality in major Chinese disasters is often wide-ranging and often in dispute. Determining mortality rates is even more difficult (see Garnault 2013). There is, unfortunately, no room here to tackle this fundamental historiographical and statistical problem, but it should certainly be borne in mind throughout the following discussion.

Older Chinese models of relief

Before charting the evolution of famine responses in modern China, it is necessary first to establish a benchmark of values and practices from earlier centuries to appreciate better the changes that are the focus of this chapter. Historically, the Chinese state presided over one of the most elaborate poor-relief and granary systems of the early modern world (Will and Wong 1991), one that in the nineteenth and twentieth centuries went through a period of severe crisis that played out in some of the largest drought- and flood-related famines in history. The north of China—large sections of which were highly unstable ecologically, such as the flood-prone and highly saline alluvial plain in northern China—was drier and generally poorer than the more-fertile and productive south, and consequently suffered many of these famines (see Figure 9.1). For this reason, this chapter focuses on northern China, using drought as a constant.

Foraging by the destitute for "famine foods" (such as leaves and tree bark) was an annual occurrence in much of China (and may even have *increased* in the twentieth century); full-fledged famine, as an event worthy of central

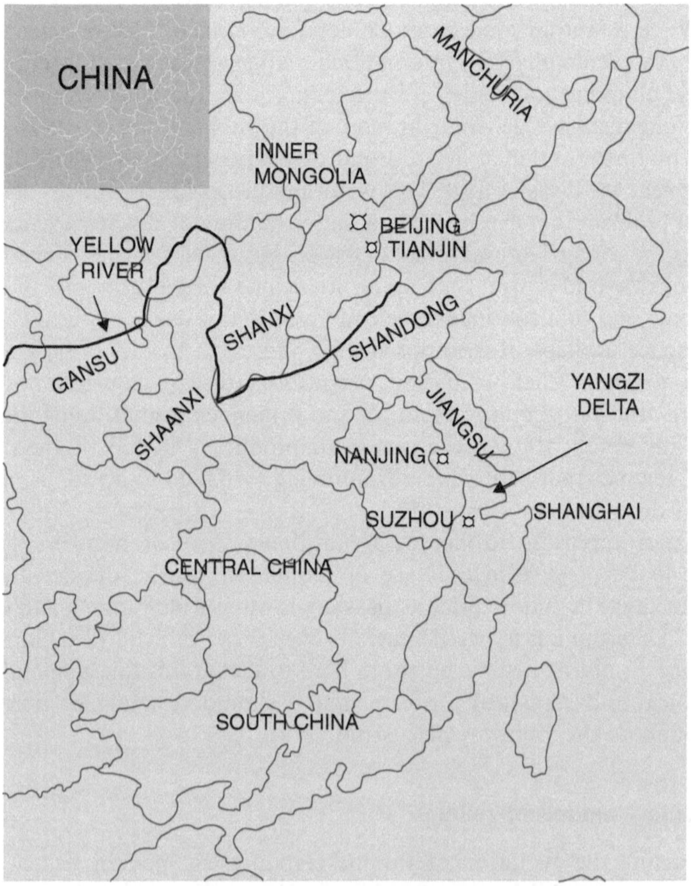

Figure 9.1 Map of China, circa 1920, showing places named in the text.
Source: Author.

government action, was then a matter of the scale of hunger and the fiscal and social disruptions that it could cause. In the three monumental pre–Mao Zedong[1] famines evaluated here, drought crippled harvests in hundreds of counties across the country's northern plains and mountainous northwest. In each famine, total food availability declined, at least at the regional level, and extraordinary food supply interventions were required to avert mass starvation.

The Qing imperial house (1644–1912), like the preceding Ming dynasty (1368–1644), was seated in northern China but drew much of its revenue from the more affluent central and southern sections of the empire. This resulted in massive volumes of annual tax grain being conveyed northward using an elaborate system of canals, and this meant that the bureaucracy centered in Beijing remained attuned to harvest levels and grain-price fluctuations around the empire. As a countermeasure to harvest or market distribution failures, the Qing presided over a centuries-old cultural repertoire of food security and famine-relief administration that peaked in effectiveness during the height of its power in the eighteenth century (Will 1990; Li 2007).

For many people, widespread crop failure rendered out of the question the main means of accessing food legally: growing, buying, or working for it. The causes of hunger are, of course, diverse and complex. For one, economic causes play out differently depending on the level of commercialization in a region, fluctuating widely even within Chinese provinces. Furthermore, while extreme poverty was certainly a factor in famines in China, where large sections of the population lived on the very margins of existence, the three pre-Maoist famines discussed here in fact occurred in regions with relatively low levels of tenancy or wealth disparity, striking, by and large, communities of poor owner-cultivators. Not surprisingly, extreme scarcity was frequently exacerbated, if not caused, by corruption, exploitation, or inaction by merchants, neighbors, or officials, leaving the desperate to turn to banditry or begging. The relief precedent established by earlier generations of Chinese, though, created expectations among the poor of a fourth means of legal sustenance—what one might term "customary relief." Expectations for this type of relief were fulfilled to varying degrees depending on time and place.

This customary relief comprised cash, clothing, coal, congee, transfers of grain (discounted, loaned, or free), and other relief goods. Considerable administration was devoted to determining who was entitled to such assistance. Imperial authorities routinely classified drought districts according to grades of harvest failure and the proportion of affected residents; through house-to-house canvassing, residents were then ranked in five grades of immediate need (Will 1990: 34). Able-bodied males were expected to seek temporary employment elsewhere during crises, or to join work-relief projects to repair canals or roads; many men routinely left for neighboring or distant regions at the first sign of harvest failure, as did farmhands, vagrants, and other residents not officially recognized as permanent members of the community (Will 1990: 257). The remaining members of "rich" and "middling" families—classifications based on criteria that changed with time and place—normally were expected to weather food crises on their own. Generally, this left women, children, the elderly, and the infirm from "poor," "very poor," and "destitute" households dependent on the care of famine-relief operations.

So far, the discussion has framed these as official measures, but in practice, the planning, financing, and management of relief operations were shared by people at multiple levels of society acting in official, quasi-official, and personal capacities. Although the ethical and religious traditions informing such efforts varied, arguably the most significant was the Confucian value system shared by the empire's gentry-elite and members of the imperial civil bureaucracy. Official and elite concern for public welfare often was couched in terms of Confucian paternalistic governance, with district magistrates and elites alike, not to mention the emperor himself, frequently positioning themselves as "father and mother officials (*fumu guan*)" of the common people (see Rowe 2001: 326–405). Extending beyond the formal state apparatus, then, relief activity sprang from the macro, meso, and micro levels. At the imperial (macro) level, a corps of civil administrators from the Qing court down to the district magistrate carried out relief provisioning and tax adjustments across the empire using an evolving body of disaster-relief treatises and manuals (Huang 1984; Will 1990). At the regional (meso) level, and mainly in more commercial areas with sufficient private resources and long-distance social networks, relief sprang from the ranks of merchant, scholar-elite, or monastic circles. And then at the district or village (micro) level, relief took the form of mutual-aid measures between farming households, and it involved handouts or congee stations set up by more capable families. These meso and micro levels of relief encompassed informal spheres of activity that worked either parallel to, or in coordination with, imperial macro measures, depending on the crisis (Zhang 2000: 96–100; Liang 2001; Smith 2009). Finally, it is important to point out that the relief systems described above apply to China's agricultural population, comprising the vast majority of Chinese until late in the twentieth century. The needy among those plying non-agricultural trades, such as artisans and peddlers and residents of commercial cities in general, were largely left to privately endowed charities such as benevolence halls (see Rankin 1986: 99–130; Rowe 1989: 100–110).

Three main points should be grasped from this:

- relief in premodern China was multifaceted and multilayered;
- relief was designed to benefit primarily farming households, which formed the core of the empire's fiscal and ideological systems; and
- relief was gendered, targeting persons deemed the weakest and most vulnerable (the old, the young, and the frail, and females in general).

The latter point is particularly noteworthy for the risks it poses to social stability, which is especially prized by Confucian ethics. While the notion of a capable "well-ordered state" with a "mandate of heaven" to rule is indeed

Table 9.1 Major modern Chinese famines

Timespan of famine conditions	*Period*	*Scope*	*Estimated mortality*
1876–79	Qing dynasty	Henan, Shaanxi, Shandong, Shanxi, Zhili (Hebei)	9.5–13 million
1920–21	Chinese Republic ("warlord era")	Henan, Shaanxi, Shandong, Shanxi, Zhili (Hebei)	500,000
1928–30	Chinese Republic (period of consolidation under the Guomindang, or Nationalist Party)	Chahar, Gansu, Hebei, Henan, Rehe, Shaanxi, Shandong, Suiyuan	Possibly 10 million
1958–62	People's Republic of China (Great Leap Forward)	21 provinces (Anhui, Guizhou, and Sichuan suffer worst)	32.5–36 million

Sources: Committee of the China Famine Relief Fund 1879:7; Xia 2000:395; Yang 2012:395–430.

central to traditional Chinese relief (Hirono 2013: S207), caring for the weakest during crises at the expense of those most likely to join the ranks of bandits—able-bodied males—suggested a paternalistic concern that went beyond merely keeping society in order. To counter these risks, communities often hired young local males as crop-watchers during periods of crisis, giving them a stake in the defense of local resources.

See Table 9.1 for an overview of casualties, damage, and costs of the four modern famines discussed in this chapter.

The North China Famine of 1876–79

A few decades after the two successive Opium War defeats to Great Britain (1839–42 and 1856–60), a calamitous drought occurred in China. The hybrid cultures and societies of the treaty ports, such as Shanghai and Tianjin, created after these defeats along the Chinese coast, were major catalysts of much of the change seen in the country's modern period. These cities would also add new dimensions to the existing disaster-relief regimes described above. Marked by a new dynamic of urban (coastal) assistance for rural (inland) communities, two major developments characterized relief of the Great North China Famine of 1876–79: new agents entered the Chinese famine field—including for the first time Western aid actors—and, largely in response to this, relief was supplied over long distances by Chinese elites who increasingly framed their efforts in nationalistic terms. In the 1870s, China's reputation in the wider world had become a motivation for disaster relief. That said, this shift cannot be taken too far: the monumental death toll of the late 1870s would largely be the result of a breakdown in imperial and local relief *capabilities*, but not in the inherited ideologies underpinning relief, sketched earlier, which were still very much part of official and rural life.

The drought in China in 1876 was possibly unprecedented in scale and severity, developing in coastal Shandong before spreading inland to four more provinces and leaving parched soil two meters deep in places. It was part of a complex global meteorological event that contributed to the deaths of between 9.5 and 13 million Chinese, as well as of millions more people in Brazil, India, and elsewhere (Committee of the China Famine Relief Fund, 1879: 7, 21; Davis 2001). In China, mountainous and landlocked Shanxi was hit hardest. After three years of back-to-back harvest failures, this once-prosperous province of banking, mining, and light industry lost, at the very least, one-third of its roughly 16 million inhabitants to famine-related death and flight.

The Qing court ordered the diversion of Beijing-bound tax grain barges to the drought zones, and then set up an official relief headquarters in Tianjin to coordinate local committees raising money from gentry and merchants, to orchestrate discounted grain sales to rein in prices that were tripling in places, and to establish distribution and refugee centers (Committee of the China Famine Relief Fund 1879: 49, 67; Bohr 1972: 30, 48, 61). Enormous obstacles, however, prevented the swift transfer of relief supplies from productive regions to the drought zone. A rain-deprived canal system largely dried up by 1877, and traffic bottlenecks in the mountain passes paralyzed grain carts heading into Shanxi (Bohr 1972: 43). Meanwhile, the Qing court was divided between assigning funds for famine relief and investing in coastal defense and in military campaigns in Inner Asia along the frontiers of the British Raj and Tsarist Russia (*Celestial Empire* 1878). The magnitude of the 1870s drought had overwhelmed an ill-prepared, under-financed, and preoccupied imperial apparatus.

Although not necessarily the first instance of missionary famine-relief in China, the size and stature of the foreign effort in the late 1870s was unprecedented, establishing orphanages and relief centers in numerous sections of the famine field, funded by 204,560 silver ounces collected worldwide, mostly from Great Britain (Committee of the China Famine Relief Fund 1879: 29–31; Hyatt 1966:110). Operating out of Christian missions that had begun sprouting up around the empire, the foreign relief activists worked separately from Chinese relief operations and were largely ignored in the famine relief pronouncements of the Qing government (Committee of the China Famine Relief Fund 1879: 25).

In contrast, China's nascent modern press centered in the treaty ports did take notice that foreigners were engaging in relief initiatives (Rankin 1986: 141). At the *regional* level, Chinese philanthropists in the Yangzi delta had already been relieving rebellion- and flood-ravaged areas since at least the mid-1800s. In the light of reports of these new foreign endeavors, though, they expanded their own charity relief operations into the distant north in 1877 and soon were working closely with official relief operations there and raising enormous sums of money (Rankin 1986: 141–150; Wue, 2004; Zhu 2006). This included 2.5 million silver ounces from a single fundraiser and Suzhou-native who also commissioned a series of graphic famine illustrations, including depictions of cannibalism that circulated in China and as far away as Great Britain (*The Graphic* 1878; Edgerton-Tarpley 2008: 131–155). Elite competition with foreign-aid efforts and a growing sense of shame at the plight of fellow Qing subjects helped to engender a national consciousness that, for the first time, entered the older mixture of motivations behind famine relief in China, at least among urban elites. Furthermore, the Qing state recognized this, taking advantage of the fundraising potential of overseas Chinese—also for the first time. Beijing appealed to Chinese in Singapore for relief monies, including from the moment it opened its first consulate there in 1877 (Peterson 2005: 94). For political reasons the state had held this group in suspicion since the fall of the preceding Ming Dynasty in 1644.

In sum, disaster relief was internationalized in 1870s China in several key ways: a precedent was set for foreign relief on Chinese soil, financed by monies from as far away as Edinburgh and Adelaide. Although the imperial government had long conducted relief operations that spanned the empire, private relief in 1870s China acquired empire-wide—"national"—and international dimensions. In other words, "inputs" (conduits for the collection, dissemination, and publicizing of human needs over the course of the crisis) increased tremendously during the famine of the 1870s, going national and global through Chinese diasporan, media, and missionary networks. But distribution "outputs" (channels for the delivery of relief on the basis of informational inputs) failed spectacularly for multiple reasons, such as inadequate transport technologies considering the scale of the crisis.

On a final note, the contributions of these media and networking innovations, while important, should not be overemphasized. Indigenous and long-standing relief practices in the 1870s likely outstripped those of Chinese urban communities or those of overseas actors: a single Shanxi county, for example, raised relief subscriptions of some 200,000 silver ounces by 1878, nearly equal to the entire foreign-aid contribution over the four years of the famine (Committee of the China Famine Relief Fund 1879: 35), and evidence of substantial village-level relief can be found in local county histories from across the famine zone (Fuller 2015). Local relief practices and inherited ideologies persisted and would continue even into the next famine under review nearly half a century later.

The North China Famine of 1920–21

The early twentieth century witnessed a steady rise in national and international charity relief efforts in China. The example of the North China Famine of 1920–21 reveals that these highly publicized national and international campaigns were crucial to the relief effort and to engendering the development of formal non-governmental relief institutions in the country. But they did not replace local, older-style relief actions. Rather, they worked alongside them. The relatively low death toll in the 1920–21 famine was as much a function of *continuities* with Qing-era customary relief measures at multiple levels of Chinese society as it was of foreign or urban innovations and interventions. In short, a multitude of relief agents and methods, including those at the village and county levels, coalesced into a flexible and resilient system in 1920–21.

By the turn of the century, the Chinese state had finally formally endorsed non-governmental relief efforts in the country. Two factors led to this development: a chastened Qing court after its disastrous backing of the Boxer Uprising in 1900, and the regime's participation in the First Hague Convention in 1899. The latter paved the way for the formal establishment of a Red Cross chapter in Shanghai in 1904 to handle refugees from the Russo–Japanese war in the northeast of China (Reeves 2005: 65–76; Zhu 2006: 488–492; Zhang 2007: 6). Soon afterward, in 1906, China's first joint Chinese–foreign relief operation formed in Shanghai to relieve a flood-related famine in nearby Jiangsu; with this event, the spearheading of foreign relief in China also passed from British to American hands (Committee of the China Famine Relief Fund 1879: 15; Darroch 1907: 3, 24).

With the collapse of the Qing dynasty in 1912, a republican government was set up, and within 9 years large sections of northern China—the same 5 provinces, in fact, that were stricken in the 1870s—experienced a rainless 12 months; between 20 and 30 million people were estimated to be destitute and in need of assistance. By 1920, the Chinese republic also was disintegrating into provincial "warlord" domains. Three crucial factors, though, helped to limit the death toll this time to some 500,000 people: (1) the hardest-hit drought areas were more accessible in 1920–21, thanks to several thousand miles of railway line built largely with foreign financing since the late Qing era; (2) the drought lasted for only a year, as compared to three back-to-back years in the 1870s; and (3) foreign involvement in famine relief in China reached an all-time high in 1921, part of a larger (mostly from the United States) outpouring of postwar relief activity in Russia and elsewhere. Urban relief committees composed of prominent Chinese and foreign diplomats and business executives from around China, along with Christian missions in the field that tapped into a global missionary network, generated and distributed close to $18 million[2] for the relief of some 7.7 million people by May 1921, including $2.48 million raised by the American Red Cross for its own work-relief projects centered in Shandong.

It must be stressed, though, that Western accounts of this famine put total relief expenditure in 1920–21 at $37 million (Peking United International Famine Relief Committee 1922: 20–26; Nathan 1965: 5–10; Li 2007: 294–302), which is an incredibly low figure for what was necessary to maintain in excess of 20 million destitute people for nine months, and it gives the impression of oversized foreign engagement of 40 percent of all relief in China that year. Estimates of required relief funds at the outset of famine conditions in 1920, however, ranged from $120–200 million (Peking United International Famine Relief Committee 1922: 16; *North China Herald* 1920). Countless unidentified Chinese relief actors were therefore left out of recognized relief totals for the famine, accounting for the huge difference between required famine funds and what histories of the event suggest was actually spent (Fuller 2013). Local and regional relief was robust in 1920, a fact little- recognized by scholars (see for example Wong 1998: 34).

During the first half of the famine in 1920, Chinese efforts at the local, provincial, and national level sustained millions of destitute people and, in many cases, continued to do so until the rains returned in the spring, and international efforts finally were fully mobilized. One major development since the 1870s was the construction of a growing railroad network in China. This was indeed instrumental in conveying relief in 1920, but the humanitarian role of technology depends, of course, on how it is used. The volume of grain that the Chinese Ministry of Communications moved down its tracks from Manchuria in January 1921, midway through the famine, amounted to a six-fold increase over previous years (*North China Herald* 1921). This is noteworthy since this southward movement of grain from Manchuria, coupled with the northward movement of some 1 million refugees to Manchuria at a cost of $7 million to the central state, fell between the domains of rival Chinese military factions that soon would be at war with each other.

An emphasis on the fluid, state-subsidized transport of relief in autumn 1920 is important for two reasons:

- It demonstrates that important members of China's "warlord" establishment acted as carryovers from Qing-era Confucian norms of paternalistic governance, initiating and overseeing, for example, mass clothing distributions and refugee shelter operations benefitting hundreds of thousands of people in greater Beijing (Fuller 2013).
- It permits the tracing of multiple conduits of Chinese relief mobilization on a national scale, which was first evident in the 1870s and which by 1920 was facilitated by a vibrant and relatively free news media that included more than a hundred daily newspapers in Beijing alone (Beijing shi difang zhi bianzuan weiyuan hui 2005: 42–53). Railway records reveal, for instance, that dozens of Chinese aid agencies—central, provincial, and county governments, as well as Buddhist groups, relief societies, and associations of sojourning merchants and workers based on hometown ties—moved 102 million kilograms of relief grain into the famine zone in December 1920 alone, much of it free of charge (*Zhengfu gongbao* 1921).

Chinese society in 1920–21 produced a wellspring of privately initiated relief at the micro, meso, and macro levels, informed by a diverse media industry that even included privately run daily and weekly periodicals dedicated to famine relief, such as the Beijing-based *Jiuzai zhoukan* and *Zhenzai ribao*, while working both parallel to and in tandem with government agencies. In other words, the overall relief effort involved a wide variety of information inputs and distribution outputs, with older and more modern methods working alongside each other and, crucially, backing each other up, including militarists and merchant-gentry acting as carryovers of imperial-era relief legacies in conjunction with international relief societies that were wiring in monies from around the world. Relief transportation was a similar hybrid of archaic and modern technologies: motor vehicles were still rarely used in relief convoys in 1920; instead, tens of thousands of oxcarts performed the crucial last leg from limited railheads to the wider countryside in more than three hundred counties.

So, what had changed? The American Red Cross utilized work-relief programs as disaster prevention measures in 1920–21, but these were hardly innovations in Chinese famine policy (Will 1990: 257–262). Instead, they were foreign interventions in an area from which the weakening Chinese state had largely de-invested since the mid-1800s. What *was* new about work relief in the republic was an emerging discourse on "work ethics" and the "usefulness" to the nation-state of the poor and famine-stricken. Recipients of customary relief in 1920–21 were still largely selected in a fashion similar to the 1870s, prioritizing females of all ages, old men, and boys in Beijing soup kitchens and village-level mutual aid programs, while excluding drug addicts and able-bodied men (Li 2007: 275; *Laifu bao* 1920; *Minyi ribao* 1920; *Yishi bao* 1920). But a discourse began to infuse the question of work relief in 1920–21 (American Red Cross 1921: 211–212), and signaled a valorization of "productivity" over unconditional assistance—what one historian has termed a "criminalization of indigence" (Chen 2012)—that would not play out fully until the re-ascendance of central state power under the Nationalist and Communist regimes.

The decade that followed the famine of 1920–21 was socially and politically transformative and, hence, by the late 1920s, famine would unfold very differently. The rise of the Nationalist and Communist parties brought a turn to Leninist-style single-party politics that soon would extend state control over this budding civil society (Fitzgerald 1996). Since the 1800s, the Chinese state had become increasingly reliant on international finance, customs revenues, and other sources of income, as opposed to taxes from its agricultural base, making it less inclined to invest in infrastructure in the ecologically fragile interior, which had been a priority of the Chinese state under the Qing. Both of these trends—valorization of "productivity" over unconditional assistance, and the objective of state strengthening at the expense of rural communities—would culminate in the Maoist famine of the 1950s. First, though, decades of civil and anti-Japanese warfare would contribute to a maelstrom of disasters from the late 1920s through to the 1940s.

The Northwest Famine of 1928–30

China circa 1930 saw a flowering of charity relief efforts that in many ways were similar to those of a decade earlier. This time, though, they were decidedly overwhelmed. In addition, fundamental changes to Chinese governance and society were well under way, sapping famine responses at all levels, from the central leadership to local community. Disentangling natural from manmade disasters is an increasingly fruitless exercise in this era. The extent of the humanitarian crises caused by droughts and floods, and exacerbated by militarism and war, from the rise of the Nationalists in 1928 to the Japanese invasion of 1937, is possibly unmatched in Chinese history up to that point (Xia 2000: 384–394; Li 2007: 307).

Non-Chinese controlled the main famine-prevention organ in China throughout most of the 1920s. The eight international relief groups that had formed in response to the famine of 1920–21 soon amalgamated into the China International Famine Relief Commission, a quasi-official agency whose constitution built in a one-man majority of foreign executives to keep the determination of relief needs and the disbursement of funds out of official Chinese hands out of a fear of corruption or mismanagement (Nathan 1965: 12). In the following years of intensified civil war, the commission devoted resources piecemeal from public fundraising drives and, from 1926 onwards, from another customs surtax, to well-digging, road construction, and credit cooperatives in various parts of the country (Xue 2008). The commission, one of a few relief agencies with national reach in the 1920s, had, in its own words, "given up" on holding annual executive meetings "due to the disorganized conditions of travel," and had not convened an annual meeting of its executive committee between March 1925 and November 1928 (China International Famine Relief Commission 1928: 11; Godement 1976: 11). Much of this disruption was because of the northward drive by the Nationalists under Chiang Kai-shek from 1926 to 1928 and the seizure by the Japanese of the main railway line in Shandong in May 1928.

The commission declared, in December 1927, that "excessive dryness" and locusts had brought "natural calamity" to 70 counties in northern China (China International Famine Relief Commission 1928: 7). By May 1929, "one of the most wide-spread and severe famines in many decades" had extended inland to the upper reaches of the Yellow River, enveloping Inner Mongolia and Gansu and Shaanxi in the northwest, where "three successive harvests … failed to materialize" in 300 counties across 8 provinces, leaving more than 50 million people in total "severely affected" (China International Famine Relief Commission 1929: 1, 1930a: 42). Yet, only an average of three grain trains a month traveled between Manchuria and northern China for much of 1928, and only 100 of 1,000 locomotives in northern China were reportedly moving at all by the following year, while great sections of the famine field doubled as a theater of civil war (Godement 1976: 90–93).

The commission's stress on meteorological factors was in part directed toward the American Red Cross, which would determine in late 1929 that "artificial causes [were] largely predominating" in the Chinese famine and so declined to join the relief effort (Bicknell 1929: 207). Internationally, the commission was largely on its own, although financed in part by a fund called China Famine Relief, Inc., based in New York City. It was also woefully unprepared: the commission had no committees in the hardest-hit northwest provinces at the beginning of 1929 because it "follow[ed], in the interest of economy, the practice of not maintaining personnel in active service in the interim between famines" (China International Famine Relief Commission 1929: 1). When it did finally project itself into the northwest, it did so minimally in the light of human needs. In one northwest province, Shaanxi, the commission brought in 408 tons of grain in 1929, handed out cash, and set up 11 soup kitchens and 9 refugee centers and orphanages when the number of famine-stricken people there hovered at 7 million (China International Famine Relief Commission 1930a: 42; Xia 2000: 388).

After assuming nominal leadership over the fractured country in 1928, Chiang Kai-shek acknowledged the importance of the international commission as a temporary entity while his regime launched its own Department of Relief Affairs that year (Janku 2011: 237, 246). It is unclear how much the Nationalist government raised from customs surtaxes and spent on relief, official pay deductions, and other means in 1928–30, but clearly it was less than its $2.3 million monthly military expenditure—one-half of its overall budget—on its attempt to unify the country under Nationalist rule (Xia and Kang 2001: 146). War, three years of drought, and the remoteness of the disaster zone combined to cause possibly 10 million deaths in 1928–30—or 20 percent of the destitute famine-afflicted population, as compared to a loss of less than 3 percent of famine sufferers nine years earlier.

Nonetheless, there are notable similarities with the dynamics of relief in 1920–21. Military Governor Zhang Xueliang, ruling over the Manchurian breadbasket after the death of his father in a Japanese bomb plot earlier in 1928, continued with the policies of dispatching relief to northern China—$1 million in this case—and settling some 300,000 refugees in his northeast domains (Janku 2011: 237–239). Military personnel also had begun to ship Manchurian grain to the drought zone on behalf of native Chinese relief agencies in autumn 1927 before the international commission had even declared a famine (China International Famine Relief Commission 1929: 8). Native-place associations produced publicity on needs in the drought zone in 1928–30 and pressured the Nanjing regime to act (Janku 2011: 251), while "purely Chinese relief organizations" from Buddhist, Confucian, and Taoist, as well as non-sectarian, circles had, the commission noted, "helped largely but inconspicuously in relief work" (China International Famine Relief Commission 1930b: 19). The Chinese Red Cross, which had played only a minor role in relief in 1920–21, made a larger contribution, opening refugee centers in Henan, Shaanxi, and elsewhere in the famine zone of 1929–30 (Li 2011: 73).

So, aside from full-fledged war, what had changed in Chinese society since 1920? At the highest levels, the Chinese state, in the form of the newly installed Nationalist regime based in Nanjing, further reoriented itself toward coastal, industrial, and urban concerns, the prioritization of state-strengthening, and the servicing of foreign debts, largely at the expense of a rural hinterland repeatedly struck by disasters (Alitto 1979; Pomeranz 1993; Thaxton 1997). Meanwhile, increased Guomindang (Nationalist Party) influence over press content as well as wartime distractions and the Great Depression (1929) reduced the public attention to the northwest famine. Poverty and refugee policy also became vested with wider concerns: usefulness to industrialization; defense against a Japanese invasion; and global stature of the nation-state (Lipkin 2006; Muscolino 2010), while the regime placed previously autonomous civic groups such as the Chinese Red Cross under state control (Reeves 2012). Many of these developments, as shown below, were extended further by the ideological rival of the Nationalists, the Chinese Communist Party (CCP), when it took power in 1949.

At the micro level, the 1930s witnessed the continued weakening of the mutual aid systems that communities had in place to combat the ecological crises that lay ahead. Official village heads, who replaced informal village leadership in the early twentieth century, now worked as tax agents for increasingly extractive military regimes that drained rural communities of able-bodied men and resources. Villages across northern China also had experienced an overall rise in social stratification and a decline in community solidarity, trends that only accelerated with the wars and disasters of the 1920s (Huang 1985: 249–274; Duara 1988; Chang 2007). Escalating civil war from 1924 made the requisitioning of farmers' carts by feuding armies standard practice across large sections of the country (Waldron 1995: 144–149), emptying entire districts of what had been an essential mode of grain transport in 1920–21. In short, while the drought of 1928–30 was geographically remote and extensive in duration and scale, it was also caused by considerably constricted relief inputs as compared to 1920–21 and outputs that either were lacking or devoted to warfare. Northern China had again experienced mass starvation akin to that of the 1870s.

Nonetheless, revolutionary fervor would soon lead to a famine that, in China's long history of disasters, was in a class of its own.

The "Great Leap" Famine of 1958–62

Alleviating hunger had been a stated goal of Chinese revolutionaries (Yue 1999: 150–183; Wemheuer 2014: 84), and the aim was often framed as an advancement of the interests of the peasant farmer through land reform. In some important ways the aims of public welfare were met in the newly established People's Republic, such as huge drops in infant mortality and gains in life expectancy. By the mid-1950s, however, in a bid for overnight industrialization (dubbed the Great Leap Forward), CCP policy stripped China's complex political economy down to a system of resource allocation by party dictate that privileged those deemed most "productive" for revolutionary and state-strengthening purposes, namely industrial workers and urban residents generally. The customary relief explored throughout this chapter so far could be seen as an extension of economist and philosopher Amartya Sen's notion of "transfer entitlements" (Sen 2010), such as social security and guaranteed employment, which are grounded in political or legal claims to sustenance, to include moral obligations that reflect normative codes of social behavior. In the mid-1950s, the CCP limited transfer entitlements to urban Chinese (in the form of the "iron rice bowl" of guaranteed sustenance from one's work unit) and enforced, mostly as a consequence on the majority rural population, new codes of behavior based on "self-reliance" and sacrifice to the state and the revolution; these codes served to delegitimize acts of relieving hunger (Brown 2012: 29–52; Wemheuer 2014: 101–114). Restrictions also were placed on migration to cities from the countryside, creating for the first time a legal distinction between urban and rural dwellers and a formal separation of urban and rural spheres (see Cohen, 1993, for intellectual background to this development). It must be borne in mind, however, that these changes were not as abrupt as they might seem, but were rather the culmination of long-term changes to Chinese political culture which, since the Nationalist rise to power and all-out war against Japanese invasion, had replaced political legitimacy predicated on basic welfare provisioning with political legitimacy based on state strength and popular sacrifice to national goals (Edgerton-Tarpley 2014). These changes, seen already in the northwest famine of 1928–30, would crystallize with the Great Leap, turning the Qing policies of resource reallocation and rural investment completely on their heads, and all at a colossal human cost.

In 1958, amid the transition to collectivization and the launch of the Great Leap, abnormal weather revisited parts of China, but certainly not at levels sufficient to account for a mega-famine across 21 provinces. The reasons for the deaths of many millions of people were complex, but almost entirely man-made and politically driven. After a brief interval of land reform and rural cooperatives, the Great Leap's communes stripped farming families of any legal claim to the grain they produced in favor of an urban industrial workforce with a prioritized food supply (Brown 2012: 53–76; Yang 2012: 168, 340). The fear and fervor of the 1957 anti-rightist movement, in which merely "talking about hunger in the countryside was considered an 'ideological problem'" (Wemheuer quoted in Brown 2012: 58), facilitated forced grain extractions from rural communities in ways that were reminiscent of the Union of Soviet Socialist Republic's attacks on the kulak farming class in the 1930s (Yang 2012: 223–229).

With the social leveling and closing off of Maoist China to much of the world, gone were charities or the possibility of international assistance; gone, too, was an even marginally independent press able to spread word of the starvation of millions, and even the age-old famine response of flight was restricted by the newly installed system of household registration (*hukou*) that tied people to their place of assigned employment (Thaxton 2008: 162–170; Yang 2012: 50, 127). The protections accorded by past regimes to farming families and females in particular were overturned, with women routinely expected to work in the commune's fields, and often despite severe malnutrition, pregnancy, or other conditions, in place of men mobilized for extensive irrigation projects (Thaxton 2008: 139–143; Hershatter 2011: 236–266). For China's rural communities, nearly all customary relief measures were gone—even punishable—pushing people to cannibalism at levels probably not seen since the 1870s (Yang 2012: 13–14, 229–233, 302).

In striking and paradoxical ways, the Great Leap took modernist critiques of the limitations of "backward" relief systems and actors in earlier Chinese famines to their logical extreme. In the obsessive search for "perfect" technical solutions to age-old problems, common sense was jettisoned in search of bumper harvests and spontaneous industrialization; attempts were made to do away with the "inefficient" family structure through communal kitchens; massive irrigation projects exhausted an overstretched rural workforce; and when mass starvation set in, the CCP's monopoly on information inputs and distribution outputs made relief impossible, even unthinkable.

Conclusion

The nature of the Chinese state and society had changed so much by the time of the Great Leap Famine of 1958–62 that viewing this event as a prototype for Chinese famines in general would be completely misguided (for an example see Becker 1996: 9–23). When the Communist state increased grain and labor levies on the rural population in the 1950s, there were no customary autonomous relief agents separate from the state, and next to no social defense mechanisms left to counter mass starvation.

Since the de-collectivization of the 1980s, rural China has experienced a steady growth in incomes. These days the CCP is more concerned with delivering a middle-class standard of living for all. And while flood, drought, and other disasters remain enormously costly to China—by one estimate, between 3–6 percent of GDP each year (Kang 2015: 23)—they are much less costly in human lives than in the past. Nonetheless, disaster relief and the elimination of subsistence crises remain central concerns to China's top leadership (State Council of the People's Republic of China 2017). Official coverage of responses to natural disasters continues to serve to buttress party legitimacy and stewardship of the country (Xinhua 2017). Interestingly, however, the forms and methods of disaster relief in contemporary China increasingly resemble those from a century ago.

Consider the droughts of 2010, the most widespread to hit China in half a century. When five southwestern provinces faced extreme drought in March of that year, followed in autumn by eight provinces in the Yellow and Huai

River basins, several factors averted a severe food crisis. China, one of the poorest nations in the world for much of the preceding century, was by 2010 flush with the cash and resources required to move massive amounts of water, grain, and other vital resources. Farmers, generally speaking, also had more savings to fall back on when disaster struck their fields. Modern infrastructure, fertilizer, and other technologies also played a role in producing food surpluses and speeding them to afflicted areas. The 2010 drought, moreover, lasted only six months.

But then central government action was also crucially important, including the transfer of millions of tons of grain reserves to drought-affected markets around the country in order to stabilize prices over the course of 2010–11 (Wang, Peng, and Li 2010). This logistical feat, as we have seen, was not in fact a function of modern infrastructure and administration, having been practiced at all levels of government in late Qing and republican famines, and with considerable success: for example, in 1920–21. The 2010 drought also saw movie stars such as Andy Lau and Jackie Chan manning phone banks to raise money from the public and from overseas Chinese. Although it may be tempting to interpret this and other outpourings of civic and private relief responses during the 2010 drought, or the Sichuan earthquake two years earlier, as Chinese society assuming modern or Western humanitarian norms—such as what followed the 2010 Haiti earthquake or Bob Geldof's Live Aid campaigns for Ethiopia in 1985—history suggests otherwise. Today's generation of Chinese global stars are merely taking up a repertoire of charity relief mobilization from a century ago when some of the greatest stage performers of the age, such as opera star Mei Lanfang, held benefit performances in Beijing and elsewhere to raise famine-relief funds. In early twentieth-century China, benefit concerts were commonplace and far-flung, from Shanghai entertainment complexes to Siberian border towns. Today's resurgence of religious, native place, and other forms of association parallels forms of organization seen in the 1920s that engendered much of that period's vibrant relief activity (Bell 2008).

The combination of these official and private relief actors in contemporary China, meanwhile, is an increasingly complex and dynamic animal that Western media coverage and commentary frequently fail to "disaggregate" into its many parts and interests (Gonzalez-Vicente 2011). The prevailing post-Mao dynamic is not one simply of state command over social institutions but, arguably, as a set of scholars has contended, one of blurred boundaries between civil society and organs of the state, a relationship in which activists play "embedded," as opposed to antagonistic, roles in official operations (Ho and Edmonds 2007). As in the past, the extent to which the Chinese party state will continue to have a dominant role in the country's reemerging civic and charity sectors, domestically and internationally, is a function of the political developments and struggles that lie ahead, and not of any essential Chinese (Confucian or otherwise) cultural inclination toward heavy-handed governance.

The appearance of a civil sector along the lines of what predated the revolution could have uncomfortable implications for Party legitimacy. The regime under Xi Jinping is now orchestrating a reversal of recent trends. These days the party considers the very idea of civil society itself a "false ideological trend" from the West (MacFarquhar 2015), while it brings the country's NGO sector back into the fold of the official apparatus (Hsu and Teets 2016). History, in such a context, becomes more acutely political. The possibility that the party may lay claim to China's disaster-relief legacies only increases the importance of further research and better public awareness today of the civic and state capacities for disaster relief in pre-revolutionary society.

In sum, the main observations made here about China have three fundamental broad implications. First, historians have put too much stress on the international dimension to disaster relief in modern China; even at their peak, foreign contributions were both considerably smaller than the totality of Chinese relief responses and slower to mobilize. Second, instances of effective famine relief in twentieth-century China were not necessarily because of the institutionalization, modernization, or "rationalization" of relief methods. On the contrary, successful relief involved social networks and relief systems already in situ at the outset of a crisis, which meshed with outside urban and foreign relief interventions in the countryside. Third, the multilayered nature of past relief in China suggests that the statist model of today's People's Republic—not to mention the Maoist model—is not an essential characteristic of Chinese social relief organization, but rather a function of a particular moment in a long history of fluctuations in state power and adaptive relief strategies. These points underscore the complexity of China's past experience of disasters, but in ways that, it is hoped, can offer lessons for addressing the challenges that lie ahead.

Acknowledgments

The author extends thanks to Georg Christ and other world history colleagues at the University of Manchester for their helpful comments and suggestions, as well as to two anonymous peer reviewers for their remarkably thorough and insightful remarks.

Notes

1 The chairman of the Chinese Communist Party from the creation of the People's Republic in 1949 until his death in 1976.

2 All dollar amounts, unless otherwise specified, are Mexican silver dollars, a common currency in China in the early 1900s.

References

Alitto, Guy S. 1979. *The Last Confucian: Liang Shu-ming and the Chinese Dilemma of Modernity*. Berkeley: University of California Press.

American Red Cross. 1921. *Report of the China Famine Relief, October 1920–September 1921*. Washington, DC: American Red Cross.

Becker, Jasper. 1996. *Hungry Ghosts: China's Secret Famine*. London: John Murray.

Beijing shi difang zhi bianzuan weiyuan hui [Beijing Municipal History Compilation Committee]. 2005. *Beijing zhi: baoye, tongxun she zhi* [History of Beijing: Newspapers and communications]. Beijing: Beijing chuban she.

Bell, Daniel A. 2008. *China's New Confucianism: Politics and Everyday Life in a Changing Society*. Princeton, NJ: Princeton University Press.

Bicknell, Ernest P. 1929. "Shall America feed China? I. No: the Red Cross Commissioners' report." *The Survey*. November 15, pp. 207–208.

Bohr, Paul Richard. 1972. *Famine in China and the Missionary: Timothy Richard as Relief Administrator and Advocate for National Reform, 1876–1884*. Cambridge, MA: Harvard University Press.

Brautigam, Deborah. 2009. *The Dragon's Gift: The Real Story of China in Africa*. Oxford: Oxford University Press.

Brown, Jeremy. 2012. *City versus Countryside in Mao's China: Negotiating the Divide*. Cambridge: Cambridge University Press.

Celestial Empire. 1878. "The imperial Chinese government loan of 1878." *Celestial Empire*, July 6.

Chen, Janet Y. 2012. *Guilty of Indigence: the Urban Poor in China, 1900–1953*. Princeton, NJ: Princeton University Press.

China International Famine Relief Commission. 1928. *Annual Report 1927 Series A No. 23*. Peking: China International Famine Relief Commission.

China International Famine Relief Commission. 1929. *Annual Report 1928 Series A No. 26*. Peiping [Beijing]: China International Famine Relief Commission.

China International Famine Relief Commission. 1930a. *Annual Report 1929 Series A No. 28*. Peiping [Beijing]: China International Famine Relief Commission.

China International Famine Relief Commission. 1930b. *Series B No. 41*. Peiping [Beijing]: China International Famine Relief Commission.

Cohen, Myron. 1993. "Cultural and political inventions in modern China: the case of the Chinese "Peasant."' *Daedalus* 122(2): 151–170.

Committee of the China Famine Relief Fund (1879) *Report of the Committee of the China Famine Relief Fund*. Shanghai: American Presbyterian Mission Press.

Darroch, John (ed.). 1907. *Report of the Central China Famine Relief Fund Committee*. Shanghai: North China Daily News and Herald Ltd.

Davis, Mike. 2001. *Late Victorian Holocausts: El Nino Famines and the Making of the Third World*. London: Verso.

Dikotter, Frank. 2011. *Mao's Great Famine: The History of China's Most Devastating Catastrophe, 1958–62*. London: Bloomsbury.

Duara, Prasenjit. 1988. *Culture, Power, and the State: Rural North China, 1900–1942*. Stanford: Stanford University Press.

Edgerton-Tarpley, Kathryn. 2008. *Tears from Iron: Cultural Responses to Famine in Nineteenth Century China*. Berkeley: University of California Press.

Edgerton-Tarpley, Kathryn Jean. 2014. "From 'Nourish the People' to 'Sacrifice for the Nation': Changing responses to disaster in Late Imperial and Modern China." *Journal of Asian Studies* 73(2): 447–469.

Fitzgerald, John. 1996. *Awakening China: Politics, Culture, and Class in the Nationalist Revolution*. Stanford: Stanford University Press.

Fuller, Pierre. 2013. "North China famine revisited: unsung native relief in the warlord era, 1920–1921." *Modern Asian Studies* 47(3): 820–850.

Fuller, Pierre. 2015. "Decentring international and institutional famine relief in late nineteenth-century China: In search of the local." *European Review of History: Revue européene d'histoire*. 22(6): 873–889.

Garnaut, Anthony. 2013. "A quantitative description of the Henan famine of 1942." *Modern Asian Studies* 47(6): 2007–2045.

Godement, François. 1976. *La Famine de 1928 à 1930 en Chine du Nord et du Centre.* Master's Thesis. Université Paris-VII, Paris.

Gonzalez-Vicente, Ruben. 2011. "The internationalization of the Chinese state." *Political Geography* 30(7): 402–411.

Graphic, The. 1878. "The famine in China—Drawings by a native artist." July 6.

Hershatter, Gail. 2011. *The Gender of Memory: Rural Women and China's Collective Past*. Berkeley: University of California Press.

Hirono, Miwa. 2013. "Three legacies of humanitarianism in China." *Disasters* 37(S2): S202–S220.

Ho, Peter. and Richard Louis Edmonds (eds). 2007. *China's Embedded Activism: Opportunities and Constraints of a Social Movement*. London: Routledge.

Hsu, Carolyn, and Jessica Teets. 2016. "Is China's new Overseas NGO Management Law sounding the death knell for civil society? Maybe not." *The Asia-Pacific Journal/Japan Focus* 14(4/3). https://apjjf.org/2016/04/Hsu.html.

Huang, Liuhong 1984. *Complete Book Concerning Happiness and Benevolence: A Manual for Local Magistrates in Seventeenth Century China*. Ed and Trans. Djang Chu. Tucson: University of Arizona Press.

Huang, Philip C. 1985. *The Peasant Economy and Social Change in North China*. Stanford: Stanford University Press.

Hyatt, Irwin T. 1966. "Protestant missions in China, 1877–1890 and the institutionalization of good works." In *American Missionaries in China: Papers from Harvard Seminars*. Ed. K. Liu, 100–120. Cambridge, MA: Harvard University Press.

Janku, Andrea. 2011. "From natural to national disaster: The Chinese famine of 1928–1930." In *Historical Disasters in Context: Science, Religion, and Politics*, 227–260. Ed. A. Janku et al. London: Routledge.

Kang, Yi. 2015. *Disaster Management in China in a Changing Era*. Heidelberg: Springer.

Laifu bao. 1920. "Pingding xian Nan'aocun cunmin Feng Sizhi jinggao gexian gecun tongbao" ("Feng Sizhi, native of Nan'ao village, Pingding County [Shanxi Province], solemnly calls on compatriots of each county and village"). October 10.

Li, Lillian M. 2007. *Fighting Famine in North China: State, Market, and Environmental Decline, 1690s–1990s*. Stanford: Stanford University Press.

Li, Hui. 2011. "Wei Zhongguo zhenwu zuoyi mofan—1928–1930 nian honghui jiuzhu xibei huabei hanzai jilue" ["Towards an example of Chinese disaster management: A sketch of Red Cross relief during the 1928–1930 drought in north and northwest China"]. In *Zhongguo hongshizi hui bainian wangshi* [*The Chinese Red Cross: A Centenary*]. Ed. Z. Chi and R. Hao, 72–73. Hefei: Hefei gongxue daxue chuban she.

Liang, Qizi. 2001. *Shishan yu jiaohua: Ming Qing cishan zuzhi* [*To Dispense Charity and Civilise: Charity Organization in the Ming and Qing Periods*]. Shijiazhuang: Hebei jiaoyu chuban she.

Lipkin, Zwia. 2006. *Useless to the State: Social Problems and Social Engineering in Nationalist Nanjing, 1927–1937*. Cambridge, MA: Harvard University Asia Center.

Liu Chang. 2007. *Peasants and Revolution in Rural China: Rural Political Change in the North China Plain and the Yangzi Delta, 1850–1949*. London: Routledge.

MacFarquhar, Roderick. 2015. "China: The Superpower of Mr. Xi." *New York Review of Books*, August 13. www.nybooks.com/articles/2015/08/13/china-superpower-mr-xi/.

Mallory, Walter. 1926. *China: Land of Famine*. New York: American Geographical Society.

Minyi ribao. 1920. "Paiyuan diaocha" ["Investigators survey soup kitchens"]. November 13.

Muscolino, Micah. 2010. "Refugees, land reclamation, and militarized landscapes in wartime China: Huanglongshan, Shaanxi, 1937–45." *Journal of Asian Studies* 69(2): 453–478.

Nathan, Andrew. 1965. *A History of the China International Famine Relief Commission*. Cambridge, MA: East Asian Research Center, Harvard University.

North China Herald. 1920. "The famine in north China: $200,000,000 needed." September 18.

North China Herald. 1921. "Manchurian grain to the famine area: how deliveries have increased." March 5.

Peking United International Famine Relief Committee. 1922. *The North China Famine of 1920–1921 with Special Reference to the West Chihli Area*. Peking: Peking United International Famine Relief Committee.

Peterson, Glen. 2005. "Overseas Chinese and merchant philanthropy in China: from culturalism to nationalism." *Journal of Chinese Overseas* 1(1): 87–109.

Pomeranz, Kenneth. 1993. *The Making of a Hinterland: State, Society and Economy in Inland North China, 1853–1937*. Berkeley: University of California Press.

Rankin, Mary Backus. 1986. *Elite Activism and Political Transformation in China: Zheijiang Province, 1868–1911*. Stanford: Stanford University Press.

Reeves, Caroline. 2005. "From Red Crosses to golden arches: China, the Red Cross, and the Hague Peace Conference, 1899–1900." In *Interactions: Transregional Perspectives in World History*. Ed. J. Bentley et al., 64–93. Honolulu: University of Hawaii Press.

Reeves, Caroline. 2012. *China, "State-centric" Humanitarianism, and the ICRC: A Historical Background.* Paper presented at the conference on "Humanitarianism: Past, Present, Future", Humanitarian and Conflict Response Institute, University of Manchester, Manchester, United Kingdom, November 8–10, 2012.

Rowe, William T 1989. *Hankow: Conflict and Community in a Chinese City, 1796–1895*. Stanford: Stanford University Press.

Rowe, William T. 2001. *Saving the World: Chen Hongmou and Elite Consciousness in Eighteenth-Century China*. Stanford: Stanford University Press.

Sen, Amartya. 2010. *Poverty and Famines: An Essay on Entitlement and Deprivation*. Oxford: Oxford University Press.

Smith, Joanna Handlin. 2009. *The Art of Doing Good: Charity in Late Ming China*. Berkeley: University of California Press.

State Council of the People's Republic of China. 2017. "China reinforces efforts in disaster relief." May 4. http://english.gov.cn/premier/news/2017/05/04/content_281475645263573.htm.

Thaxton, Ralph. 1997. *Salt of the Earth: The Political Origins of Peasant Protest and Communist Revolution in China*. Berkeley: University of California Press.

Thaxton, Ralph. 2008. *Catastrophe and Contention in Rural China: Mao's Great Leap Forward Famine and the Origins of Righteous Resistance in Da Fo Village*. Cambridge: Cambridge University Press.

Waldron, Arthur. 1995. *From War to Nationalism: China's Turning Point, 1924–25*. Cambridge: Cambridge University Press.

Wang, Huazhong, Peng Yining, and Li Yingqing. 2010. "Dry spell hits where it hurts." *China Daily*, April 9. www.chinadaily.com.cn/china/2010-04/09/content_9705016.htm.

Wemheuer, Felix. 2014. *Famine Politics in Maoist China and the Soviet Union*. New Haven: Yale University Press.

Will, Pierre-Étienne. 1990. *Bureaucracy and Famine in Eighteenth-Century China*. Stanford: Stanford University Press.

Will, Pierre-Étienne., and R. Bin. Wong. 1991. *Nourish the People: The State Civilian Granary System in China, 1650–1850*. Ann Arbor: University of Michigan Press.

Wong, Linda. 1998. *Marginalization and Social Welfare in China*. London: Routledge.

Wue, Roberta. 2004. "The profits of philanthropy: Relief aid, Shenbao, and the art world in later nineteenth-century Shanghai." *Late Imperial China* 25(1): 187–211.

Xia, Mingfang. 2000. *Minguo shiqi ziran zaihai yu xiangcun shehui* [Natural Disasters and Rural Society in the Republican Era]. Beijing: Zhonghua shuju.

Xia, Mingfang., and Kang Peizhu. 2001. *20 shiji Zhongguo zai biantu zhi* [Chinese Disasters of the Twentieth Century]. Vol. 1. Fuzhou: Fujian jiaoyu chubanshe.

Xinhua. 2017. "China Focus: Quake rescue demonstrates China's strength." August 10. www.xinhuanet.com/english/2017-08/10/c_136513368.htm.

Yang, Jisheng. 2012. *Tombstone: The Untold Story of Mao's Great Famine*. London: Allen Lane.

Yishi bao. 1920. "Guanban zhouchang" ["Officially-run soup kitchens"]. December 7.

Yue Gang. 1999. *The Mouth that Begs: Hunger, Cannibalism, and the Politics of Eating in Modern China*. Durham, NC: Duke University Press.

Zhang, Jianqiu. 2007. *Zhongguo hongshizi hui chuqi fazhan zhi yanjiu* [*A Study of the Beginnings and Development of the Chinese Red Cross*]. Beijing: Zhongguo shuju.

Zhang, Xin. 2000. *Social Transformation in Modern China: The State and Local Elites in Henan, 1900–1937*. Cambridge: Cambridge University Press.

Zhengfu gongbao. 1921. "Jiaotongbu mianfei yunzhen pingzhao jishubiao—jiu nian shi'er yue fen" ["Accounting chart of relief transported free-of-charge by the Ministry of Communications during December 1920"]. June 13.

Zhu, Hu. 2006. *Difangxing liudong ji qi chaoyue: Wan Qing yizhen yu jindai Zhongguo de xinchen* [Localist Circulation and its Transcendence: Late Qing Charity Relief and Modern China's Supersession of the Old by the New]. Beijing: Zhongguo renmin daxue chuban she.

10 Bitter greens and sweet potatoes

Food practice and memories of hunger in rural China

Ellen Oxfeld

Introduction

When making connections between food and memory, some have argued that industrialized food consumption creates mass amnesia. For instance, Sidney Mintz compared modern food consumption in North America to "the hunting-and-gathering existence of our species, when food was eaten as it became available, without much reference to situation or circumstance"; he concluded that when food is consumed in such unstructured non-events, then both the food and the event are quickly forgotten (Mintz 1985). For Mintz, phenomena like fast food and manufactured food create uniformity over many different spaces and time zones and turn eating into a process of "refueling" emptied of local significance.

Eating events unrelated to specific times and places are just one illustration of the relationship between time and space in modernity as characterized by Anthony Giddens. Modern time, states Giddens, is "empty time" (Giddens 1990: 17) because it is not marked by festivals, customs, and practices indicative of time in a specific locale (Giddens 1990: 9). The notion of "disembedding"—the sense that modern life separates time from particular spaces, and has ripped apart the connections, or the "embedding," of social activities in particular contexts—is critical here (Giddens 1990: 20). In the context of food and diet, such "empty time" might be contrasted with what David Sutton describes as time marked by "local culinary traditions" that "reenact belonging and identity at every meal"; such culinary traditions, says Sutton, are part of "cultural memory," because "one cannot enact what one cannot remember" (Sutton 2001: 8).

In real life, of course, societies are not dichotomized quite so simply as to separate those in which local culinary traditions are firmly embedded (and thus closely tied to cultural memory) from those with industrialized food emptied of all meaning. Symbolically meaningful eating practices are not entirely absent from societies with industrialized food systems and a predilection for fast food; such practices may be tied to local traditions, such as a regional festival, or to national ones, such as Thanksgiving in the United States.

Nonetheless, this analytical distinction between local culinary traditions (those rich in cultural memory) and industrialized food (characterized as lacking layers of meaning and memories) is a useful way to begin thinking about the connections between food and memory in Moonshadow Pond, a village where I have been engaged in field research periodically since the early 1990s in rural Meixian, a Hakka county in Guangdong, China.

In this chapter I investigate the ways in which particular food substances and food practices in Moonshadow Pond embody historical memory and become markers of the historical transformation from food shortage to food abundance. My focus will be on the way villagers find significance in the past and, indeed, remember the past, through their discourse about these foodstuffs and food practices. The meanings of these foodstuffs are not merely fixed in an overarching semiotic system; rather, their significance has changed over time as they have interacted with historical events and transformations. Furthermore, village residents often view the historical changes themselves through the lens of food.

In thinking of the food/memory nexus, we must consider what Jon Holtzman has characterized as the unconscious elements of memory with regard to food. As he reminds us, many elements of food culture are embodied in "practices that may, or may not be clearly reflected in conscious memory" (Holtzman 2009: 41). We can understand these practices most cogently through the use of Bourdieu's notion of "habitus"—internalized and taken-for-granted practices related to one's class position, or to other elements of one's social identity, in a given social context. As such, some elements of memory can be "verbally narrativized" whereas other elements of memory "are rooted in unreflected-upon bodily practices" (Holtzman 2009: 41).

Examples of verbally articulated memories are numerous, and they include the ways certain foods prompt people to think about and speak about the past. For instance, literature on immigration often notes how growing (or purchasing, cooking, and consuming) foods that remind the immigrants of home can forge important connections to their country and culture of origin (Williams-Forson 2014; Ray 2001). Recollections of the foods and meals of home are especially evocative for refugees who are unable to return home (Song 2012).[1] In her work on postwar memories in Sierra Leone, Susan Shepler explains that food is often central to narratives of war, in part because although war contains many different kinds of atrocities, food ties individual experiences to larger processes and structures, and the experience of food "evokes recollections that are not simply cognitive, but also emotional and physical" (Shepler 2011: 45). As such, after the war's end, many recollected their experiences as civilians or combatants through their different experiences of food.

Ralph Thaxton uses the term "embodied memories" to further categorize a relation between memory and food. Examples include damage to the body from past food shortages or chronic stomach problems caused by having consumed coarse ingredients to stem one's hunger during a famine. Thaxton juxtaposes such embodied memories with consciously articulated "semantic memories" (Thaxton 2008: 302–304). My landlady, for instance, used to speak about her bad digestion as stemming from the coarse food she ate during the Great Leap Forward famine, such as rice chaff ground up and made into steamed buns. We now know that exposure to the Great Leap Forward famine during fetal life or early childhood has been linked to greater risk of hyperglycemia during adulthood, especially once the adult is living in a more nutritionally rich environment (Yanping Li et al. 2010).

But even when memory is consciously articulated, we need to keep in mind that food, even in sparking conscious memory, often operates at unconscious levels because of its power to affect our senses. Sutton uses the word "synesthesia" to invoke food's power to arouse memory. As he explains it, "Food's memory power derives in part from synesthesia, ... the synthesis or crossing of experiences from different sensory registers (i.e., taste, smell, hearing). Taste and smell ... become evocative of social situations with which they are associated" (Sutton 2001: 17).

With these caveats in mind, our main focus in this chapter will be on the conscious elements of memory articulated by individuals rather than those unconscious ones attributed to bodily constitution or predisposition to disease. We will also need to keep in mind that as in any place, memories in Moonshadow Pond will differ in terms of individual and generational experiences. Foods such as the sweet potato, which evoke memories of lean times for older and middle-aged people, may have no such impact on younger people who did not live through those times themselves, but at most receive these memories second-hand when they hear stories from their elders.

Contrary to what one might assume, not all memories of food during the collective era are associated with scarcity and bad times. Food, after all, can also elicit nostalgia. Villagers' discussions of food during the collective era are thus not without contradictions. In recalling their memories of this time, people may talk about the Great Leap famine or the scarcity of food throughout the collective era in general, but they may simultaneously recount wistfully how meals and mealtimes were characterized by a greater sense of social solidarity than in the present. For instance, Ailing, a middle-aged woman who was an adolescent in the 1960s, remembered the communal nature of collective-era mealtimes. She recalled how everyone piled food on top of their rice or congee (rice porridge), took their bowls outside, and ate together perched on their knees, on door thresholds, or sitting on benches. As she further recollected, the food was not abundant, but the experience was a shared one. "Not like now," she griped, "when everyone is just inside watching television."

I begin my analysis in the next section with a discussion of how the relationship of food to memory pertains to holidays and lifecycle rituals. I then turn to a discussion of several common foodstuffs and the ways in which they not only signal a consciousness of historical transformations but also provide a framework against which to think about them. By concentrating on how food summons memories and links people to the past, whether by reminding them of disruptions and discontinuities or of positive communal associations with that past, we can begin to think of food in Moonshadow Pond as part of an overarching semiotic system. But it is certainly not a static one.

Memory, food, and calendrical rituals

Moonshadow Pond residents say that the three most grand (*longzhong*, 隆重) annual holidays are the Lunar New Year, the Mid-Autumn Festival, and *zuofu* (a local celebration related to the Ghost Festival, but not overlapping it). These celebrations involve large family meals or banqueting and interfamily visiting. Each sparks a variety of memories and discourses about the past, some of which are contradictory or ambivalent (see Holtzman 2009), and many of which are triggered by the role of food. Furthermore, as China has moved from the collective to the reform era, changes in national policies have been inextricably linked with people's memories of these holidays in either overt or more subtle ways.

On the one hand, the festivities of the present are often a springboard to memories of past scarcity and hardship for those who lived through these changes. But another, more nostalgic narrative bemoans the gradual loss of some foods or food customs because of peoples' busy lives, or because substitutes for food once laboriously produced at home can now be found in the market. It is important to point out that the "past scarcity narrative" and the "nostalgia for old foods narrative" provide a way to talk about the present by contrasting it with the past—that is, to see both past and present through the lens of food—but they are not to be taken at face value as historical accounts.

For instance, Teacher Liu's recounting of food's connection to calendrical festivals was typical of those narratives in which holidays became a springboard for talking about the relative food scarcity of bygone times. "In the past, we never had much to eat," Teacher Liu wrote to me in a letter in 2007. Teacher Liu had spent many years teaching in the primary school in Moonshadow Pond and was in her early thirties when she wrote this. She was old enough to remember times of relatively less abundance, having grown up in the early 1980s, but certainly she was not old enough to have personally remembered the most trying times. And yet her focus on holidays allowed her to shed light on relatively less-secure times—either those that she experienced herself, or those she heard others speak about. As she put it,

> We usually just relied on our own vegetables, and we lived frugally (*sheng chi jian yong,* 省吃俭用). It was only during holidays that we would get meat, make some *ban* (treats such as the *nianban*, made from either glutinous rice flour, or wheat flour), and comfort our stomachs. Therefore, until the present, "what holiday" means "what will we eat?"

"When we were growing up," Teacher Liu explained, "we always used to look forward to the Lunar New Year and to the time when we would have *nianban*. The children loved to eat them, but now the children don't want them."

Baoli, whose family was my host in Moonshadow Pond, also reminisced about the past. He then continued with a complaint about the packages of factory-made cookies that his grandson was so fond of.

One elderly woman in Moonshadow Pond used her memory of holidays to remark on culinary change. She noted that holidays of the past were celebrated in much less grandiose fashion when it came to food. Comparing the rather elaborate Mid-Autumn Festival meals of the present with the simple celebrations remembered from the past, she said,

> All we used to do during the Mid-Autumn Festival was to make some deep-fried taro balls (*jian yuyuan*, 煎芋圆) and savory deep-fried wheat thins (*sanzi*, 馓子) and we would also eat peanuts and moon cakes. In the evening we would worship the deities and wait for the moon's sister to come down and eat the moon cake. The kids would play with the vegetable basket and put a man's clothes in it. They would ask the moon's sister to come down and if the clothes moved we would be really scared and all run away! This happened in the old society and in the collective era as well. Now these things like candies and peanuts and even moon cakes are not that special for kids anymore.

It is interesting, especially considering Baoli's remarks about the declining appreciation for *nianban*, that this villager's reminiscence reflected less on a celebration of contemporary abundance and more on the inability of today's youth to appreciate the simpler treats of the past. Indeed, Baoli once said that the work involved in making *nianban* today was not equivalent to the work they required in the past. Baoli noted that children used to love and anticipate *nianban*. But now, with their desire for factory-made cookies, "We make many fewer *nianban* than we used to." In this case, he looked back nostalgically and used his memories to critique present-day trends.

Despite Baoli's complaints about the current lack of appreciation for traditional treats, much about the contemporary celebration of the Lunar New Year in Moonshadow Pond does appear connected to tradition. The preparation of traditional foods such as *nianban*, braised ducks, and sausages still occupies people for days. Food is offered as a kind of prestation, or obligation, to ancestors and local deities as well. In each house, on the morning before the New Year, offerings of food are presented to heaven, the Stove God, and to one's ancestors; villagers then carry these offerings in baskets strung from bamboo poles balanced over their shoulders. They walk to the village entrance, where they lay out their offering to the village guards,[2] and then proceed to the lineage temple and lineage branch temples to honor their ancestors, laying the same offerings out again. These prestations include the “three animal sacrifices” (*san sheng*), chicken, pork, and fish, as well as vegetarian offerings (*zhai pan*)—including *nianban*, citrus fruits, apples, and candies—and finally tea and glutinous rice wine.

On New Year’s Eve, the family sits together for *tuanyuan fan* (团圆饭), meaning a unity or reunion meal. Then, at midnight, the God of Wealth (*caishen*, 财神) is welcomed at the front door. The next morning, the family eats breakfast together in another reaffirmation of unity.

Throughout the holiday, food takes on a greater symbolism than it does in everyday meals with the focus placed on “food homonyms” for prosperity, harmony, and family unity, just as certain foods are significant on other auspicious occasions such as marriages, house movings, and new business openings. For instance, the reunion meal should include fish as the word for fish (*yu*, 裕) is a homonym for the word for “plentiful,” and therefore connotes prosperity or abundance. On New Year’s morning, celery is added to the morning congee because the word for celery (*qin*, 芹菜) has the same sound as the word for industrious and hard-working, and one needs to work hard to prosper. Also, garlic stems are stir fried that same morning because the word for garlic (*suan*, 蒜) sounds like the word for calculate, and in order to prosper one most certainly needs to calculate properly! Many of the *nianban* are round, to emphasize unity. Further, many of them contain homonyms for other words that mean “to prosper.”

The two weeks of the Lunar New Year holiday include visits from (and meals with) married daughters and their families, and much back and forth movement occurs as friends and relatives visit each others’ homes and are plied with tea, *nianban*, biscuits, pomelos, peanuts, and other treats. Nonetheless, villagers often reflect on these holiday practices as something new, as they never before had the material abundance to celebrate the New Year with such a cornucopia of foods.

A few villagers also pointed out to me that this contemporary feasting was not only different because of the profusion of foods, but also because its meaning had changed. During more radical periods of the collective era, the government attempted to introduce new symbols for the New Year, particularly regarding the family unity meal. These “new” symbols were jettisoned during the reform era for more traditional meanings. For instance, although no one would admit to practicing it themselves, a number of people told me about the “recall bitterness meal” (*yi ku fan*, 忆苦饭) that the authorities during the Cultural Revolution (1966–69) had urged them to eat before consuming their family unity meal. The idea was to boil wild vegetables together with rice chaff and make a paste—certainly an extremely unappetizing gruel. Each family member was supposed to be given a bowl of this “recall bitterness” dish, and while eating it they would be prompted to recall the “bitterness” of their past lives in the preliberation era, referred to as the “old society” (*jiu shehui*, 旧社会). These memories were also supposed to prompt thoughts about and gratitude for the relative “sweetness” of the present. This was all prompted by Mao’s call to “recall the bitter before thinking of the sweet” (*yiku sitian*, 忆苦思甜). The aim, of course, was to build support for Communist rule and especially for the radical path of the Cultural Revolution, and the use of food to spark memories of the bitter past of the old society was one way to do this. (Thinking back to the old society was also supposed to erase memories of the starvation of the Great Leap Forward.) But a number of older villagers told me that while this practice *did* make them remember the "bitter," it was not the bitter of the old society, but precisely the harsh times during the Great Leap Forward (1958–1961) when they were reduced to surviving on such things as buns made from rice chaff.[3]

Food as embodied memory

I now turn from the issue of memories brought forth by food on particular festival days to examining the ways in which specific foods have come to embody memories. These foods "stand for" particular elements of the past for many residents of Moonshadow Pond. At the same time, these food items may now be consumed in new contexts, and they may not necessarily index the same meanings to younger people unaware of their former uses and settings. Although numerous foods in Moonshadow Pond embody powerful memories of the past, I focus specifically on three: bitter greens, sweet potatoes, and congee.

Kumai ***and other ordinary vegetables***

A ubiquitous leafy green vegetable, the "bitter veined" vegetable kumai (*kumai cai*, 苦脉菜), is available from winter through August.[4] As implied by its name, it has a bitter taste, particularly when boiled or steamed. Villagers say that during the old society and collective era it was especially "bitter" tasting because people had little or no oil to cook with, which could have improved its flavor.

With prosperity, however, came an ever-expanding number of vegetables that were available for growing in Moonshadow Pond. Indeed, according to the *Meixian Gazetteer*, the number of commonly grown varieties has increased from between 20 to 30 (at the time of Liberation) to between 70 and 80 now (Meixian Zhi 1994: 269). Such a great increase in variety suggests that there is no longer a need to be nostalgic about vegetables from the past, and indeed villagers do not articulate nostalgia for their former economic hardships and limited diet. But people sometimes do express nostalgia for what they portray as the simpler and less corrupt social relationships of the past, as well as the greater sense of security those relationships fostered. As such, kumai cai, although associated with the greater poverty of both the old society and the collective era as compared to the reform era also has a number of positive connotations.

For instance, villagers like to tell folk stories about kumai cai that show how, despite its bitter taste, it has high nutritional value, whereas higher-status vegetables with sweeter tastes may in fact be nutritionally worthless. In one such story, a man with an adopted son and a son by birth decided to discriminate against the adopted son by feeding him the lower status vegetable kumai *cai*. In contrast, he fed the son by birth a sweet-tasting green (amaranth, *xian cai*苋菜, sometimes translated as "Chinese spinach"). The father thought he was favoring his son by birth by giving him the tastier green, but instead the adopted son grew strong with the higher nutritional value of the kumai cai while the birth son grew weak.

Kumai is also traditionally mixed with congee (rice porridge). Since congee contains a great deal more water than regular rice, it enables the rice to last longer, and thus villagers consumed it much more frequently prior to the reform period. While congee is still a common breakfast dish in Moonshadow Pond, cooked rice rather than rice porridge is now the more common staple for lunch and dinner. *Kumai kao zhou* (苦脉靠粥), which means that kumai accompanies congee, is another common saying often quoted to me, indicating that kumai's relationship with congee is an aspect of its ordinariness as well as its appeal.

Of course, kumai, like many other foods, can be used in less mundane ways, especially in restaurants when mixed with foraged foods once more widely available, but which in the reform era have become high-priced specialty items. In this sense, kumai's meanings have changed along with the economic transformations from the old society, through the collective and reform eras. An example of this is an eel that can be stir-fried with kumai cai. The *huangshan* (黄鳝) is a swamp eel that used to live in abundance in flooded rice paddies. It was originally mixed with kumai cai to counter the green's "bitter" flavor and improve its taste. Baoli remembered gathering swamp eel—especially during the Great Leap Forward, when hunger was widespread—because it was a significant source of protein. Elders also spoke of their memories of mixing huangshan and kumai cai during the old society. As pesticide use in rice paddies has become more widespread, however, huangshan have become more rare (although one can still occasionally find them in the flooded paddies). Now, swamp eel have made a new appearance in restaurants, and huangshan stir-fried with kumai has become a banquet specialty.[5] While kumai cai reminds old people of the "bitter" adversities of the past as well as the era's simple pleasures, the vegetable represents something different to younger generations who now experience it as a pricey dish in restaurants.

Sweet potatoes

When elders speak about the past, and about the ways their lives have changed, they usually come back to this frequent refrain: "All we ate was sweet potatoes." The sweet potato is called *fanshu* (番薯) or "foreign potato" (in Hakka and many other Chinese dialects), presumably because it was introduced to southeast China from abroad (Wang 1995: 11). The governor of Fujian Province has been credited with importing the sweet potato in 1594 from the

Philippines, where there was already a Fujianese community, in order to give local peasants a staple they could subsist on in the case their grain crops failed (Simmonds 1999: 102; Spence 1977: 262). By the late eighteenth century, the sweet potato had "become a staple food for the poor in the coastal southeastern provinces" (Spence 1977: 263). The tuber spread rapidly to other parts of China as well, and even in the north it became an important part of the diet, allowing a larger population to subsist than would have been possible without it (Li 2007: 109, 313).

Although farmers always tried to sell some of their grain crops, they grew most sweet potatoes for immediate consumption because they were perishable. John Lossing Buck notes in his 1937 study of farm families in China that those in the double cropping rice area of China consumed up to 92 percent of the sweet potatoes they grew, as opposed to 66 percent of the grains that they produced (Buck 1937: 401). Martin Yang's 1945 ethnography of the village of Taitou in Shandong Province vividly illustrates the widespread use of the sweet potato to stave off hunger in rural China. Yang categorized the entire village population based on food consumption, and identified the very poorest as those "for whom sweet potatoes are the main item of the diet … at every meal every day throughout the year" (Yang 1945: 32).

The sweet potato is also an advantageous crop because it grows on "dry land," and thus does not use up paddy land. Being an extremely hardy plant, it can grow in adverse conditions. Ordinary people in the Meixian area consider the sweet potato to be among several "miscellaneous foods," or provisions (*zaliang*, 杂粮) (Fang 2005: 9).[6] In the past, however, its versatility and dependability made it an important staple that could be prepared in many different ways. Sweet potatoes were boiled, steamed, or roasted, cut up and added to congee, dried, added to soup, and turned into starch for making steamed buns or as a thickening agent in cooking. In the past, people were especially dependent on sweet potatoes when they had run out of the first crop of rice but had not yet harvested the second crop. Additionally, sweet potato leaves could be stir-fried and sweet potato stems could be fed to pigs.

Because they were filling and common, sweet potatoes figure prominently in recollections about food scarcity, regardless of whether the individual talking about them is remembering the old society or the collective era. For instance, in 2007 a school teacher in Moonshadow Pond compared his childhood memories from the 1960s to his son's present-day experience: "My son doesn't even know the meaning of being hungry," he complained one night. "At that time, all we ate was sweet potato, and the starch of the cassava (*shufanshu*, 树番薯) mixed with water to make us feel full. If we had even a bit of pork we thought it was a special treat." When I asked older people what they ate during childhood, they almost always included sweet potatoes in their memories. Many simply said to me, "All we ate was sweet potatoes!" Others also mentioned rice gruel as an accompaniment for sweet potatoes. One old woman told me that when her son was born in 1948, all she gave him to eat after weaning was watery congee and sweet potato.

The sweet potato also figures in published accounts and memoirs of the Great Leap Forward famine, which refer to it as a last-resort survival food. For instance, in Yang Jisheng's monumental study of the famine, he quotes a party secretary from a prefecture in Henan who remembered how, in Sichuan, people were so hungry that they ate sweet potatoes as they pulled them from the soil (Yang 2008:210). In Moonshadow Pond, many people remember that areas where the food situation was not as bad as others during the Great Leap Forward, such as the Chaozhou area of Guangdong, were often where the sweet potato was still available, even if nothing else was. It is important to point out that sweet potatoes are, in fact, quite nutritious, and where available in sufficient quantity, undoubtedly saved villagers from starvation. But the fact that it was a "last resort" when nothing else was available is one reason its memory is linked so heavily to food shortage and even starvation.[7]

The sweet potato also had a decidedly "rural" connotation. When I traveled to Hong Kong with some dried sweet potatoes, some of my friends commented that sweet potatoes were something people in the countryside ate. One of my neighbors in the village, Red Chong, told me a story about Ruolan, my Hong Kong friend who introduced me to Moonshadow Pond. Although her family had roots in Moonshadow Pond, they had lived in India and only returned to China after the Sino-Indian conflict of 1962, when many Indian Chinese were deported. Ruolan's family then settled in Guangzhou. But during the Cultural Revolution, when Ruolan was required to move to a rural area, she chose Moonshadow Pond because of her kinship connections there. She stayed for four years before finally immigrating to Hong Kong (where her husband already resided). As Red Chong told me, "When Ruolan was in the village, her husband would send her things from Hong Kong, which she considered more nutritious. She thought that sweet potatoes were not good for you, and could not be digested well, so she would not let her son, who was about four then, eat them. He loved them, though, and would steal them when she wasn't looking!"

The neighbor's story may suggest that someone (like Ruolan) with an urban background cannot appreciate the value of a simple foodstuff; it brings to mind as well the parable of the man who, based on a presumption, chose to feed his adopted son "common" bitter greens rather than the "special" amaranth he reserved for his natural son. The neighbor's story aligns him positively with his rural background through his appreciation of the sweet potato, but rural speakers can also use the "sweet potato" pejoratively. To call someone a "big sweet potato" (Fang 2005: 10) is to speak about them as too simple or even stupid.

Indeed, the very "rural" connotation of the sweet potato, as well as its more perishable nature (compared to grains), may be why it figures so importantly in the memories of older village residents as a hunger killer. As Frank Dikötter tells us in his study of the Great Leap, there was at least one reason why the state declined to procure sweet potatoes from the rural areas during that time: urban residents regarded them as peasant food and had little interest in eating them. Of course, as I mentioned earlier, there was a practical reason for not providing sweet potatoes to city dwellers, since they spoil much more rapidly than grains (Dikötter 2010: 136). Therefore, any grains produced during the famine were often allocated for the cities, whereas sweet potatoes alone remained for villagers to eat. As a result, "cadres responded to pressure to increase the yield by switching to the tuber, which was easy to cultivate. More often than not, farmers were left with potatoes only" (Dikötter 2010: 136).

During the reform period, the fate of the sweet potato in Moonshadow Pond has also changed. With the abundance of meat, fruit, and numerous varieties of vegetables, the sweet potato is no longer necessary to create the feeling of fullness, and villagers no longer mix it with congee or make it into flour. Indeed, the *Meixian Gazetter* reports that the amount of land dedicated to growing sweet potatoes in Meixian decreased by two-thirds between 1955 and 1987 (Meixian Zhi 1994: 265). Now, sweet potatoes are often used for pig feed. But sweet potatoes still function as a snack, boiled and consumed between meals, particularly for children. Yinzhao, a village woman who worked actively cultivating rice and vegetables into her late seventies, summarized the drastic change in the role of the sweet potato by saying that she grows only a few sweet potatoes now, "because people only eat one or two, and then no one eats the rest."

When cooking the sweet potato now, villagers combine it with expensive ingredients that would not have been possible to use during the lean years. For instance, sweet potato, ginger, and sugar are combined to make a soup to calm stomach distress. One day when my stomach was upset, Baoli's wife, Songling, made some of this soup for me. She immediately noted that in the past the sugar would have made it too expensive to use sweet potatoes in this way. Many of Moonshadow Pond's older residents now contrast the sweet potato with new vegetables better suited to more prosperous times, especially greens such as spinach, cilantro, or green cabbage (*bao cai*). Cabbage, for instance, needs to be cooked with lots more oil than sweet potatoes require, and spinach and cilantro provide less satisfaction than the stomach-filling sweet potato. Interestingly, I heard some village residents speculate that sweet potato may one day become a prestigious heritage food like other commons foods of the past. As fewer farmers plant the sweet potato, and as it becomes a less important part of the diet, its rarity will cause its prestige value to rise. Some speculate that it will one day become a special culinary item in the banquet halls.

Congee versus rice

Similar to sweet potatoes, congee (rice porridge) is associated with the past in the minds of older villagers. Congee can be served as a thick porridge or watered down to a thin gruel with almost no nutritional value. Villagers frequently recount memories of trying to subsist on such a watery gruel in their stories about the collective canteens during the Great Leap Forward. As mentioned above, congee, as opposed to dry rice, is associated with the collective era in general. I heard a number of stories about trying to subsist in the past on a diet of watered-down congee paired with sweet potatoes or dried vegetables. Congee contrasts with rice in these stories—with dry rice described as a luxury of the reform era.

But, unlike the sweet potato, congee is still a main feature in the diets of most villagers; they may no longer subsist on watery rice gruel, but they do eat congee almost every day, usually for breakfast. Congee can also be prepared as an extremely tasty and savory dish with the addition of items such as pork, dried mushrooms, celery, squid, dried radish, or greens. In other areas of Guangdong, in fact, such rich concoctions of congee are actually savored, and in the Chaoshan area it is favored so much that it is even served at banquets (Chen 2001: 20).

Because of its elasticity and use in the current diet, congee sparks different associations in peoples' minds than do sweet potatoes, and it does not always connote the scarcity of the past. Congee is sometimes thought to be ideal for the very old and the very young because it is easier on the digestive system than dry rice. For instance, today's grandparents often take care of their grandchildren because the middle generation is out working. When I asked these grandparents what they fed their grandchildren, I would inevitably get a mixture of answers. Some said they saw no necessity for the relatively newer trend of giving children milk, either powdered milk or the fresh goat milk available in the village. Instead, they opted for the traditional *geng*, a porridge made from rice flour. Or, they would give slightly older toddlers congee, with other items, typically greens, which were added gradually as the child was able to digest them. Some elders said they did prepare powdered milk for their grandchildren as a special treat, but that *geng* and congee with leafy green vegetables were the main staples for a toddler's diet. (They also commented that because of the scandals of recent years involving adulterated milk powder, they would only buy "foreign" milk powder if they used it, although, in fact, some of the scandals have also involved foreign companies.)

Considering those factors and congee's continuing role in the contemporary diet, congee did not solely elicit the images most often associated with the sweet potato—that is, images suggesting a past of poverty connected to rural identity. Still, the contrast between congee versus rice is still useful to evoke one difference between past and present: poverty versus relative plenty.

Conclusion

If we understand memory as "the notion of experience or meaning in reference to the past" (Holtzman 2009:363), then bitter greens, sweet potatoes, and banquets that celebrate calendrical rituals (such as New Year) can be said to represent historicized foods or food experiences in Moonshadow

Pond. Residents there do not understand these items solely in terms of the here and now, but rather with reference to historical changes and social transformations. As such, they are a lens through which to access historical consciousness in rural China. But this historical consciousness is necessarily more acute for those who actually experienced the old society, or at least have some memories of the collective era. Young people did not, as a rule, tell me stories about sweet potatoes or kumai cai, as these foods did not evoke any particular memories for those born during the reform era.

Notes

1 Weijie Song analyzes the voluminous writings of the Chinese author Liang Shiqiu (1903–1987), who left China in 1949 and spent the remainder of his life in Taiwan. He wrote numerous small pieces describing the many regional cuisines of China, and especially the food of Beijing that he left behind. As Song describes, these writings helped to "feed … the diasporic mind craving an imagined homecoming to a distant, idealized Beijing" (Song 2012: 89). For Liang, it was "human nature to maintain a nostalgic longing for the distinctive culinary flavor of one's hometown" (Song 2012: 98). His "nostalgic sentiments aim not only at the flavor and taste of the lost hometown cuisine but also at an irretrievable mode of life and culture that has disappeared along with a historical epoch" (Song 2012: 103).

2 Sheguan Laoye社官老爷and Gongwang公王

3 Ralph Thaxton points out that in Da Fo Village, the subject of his study of the Great Leap Famine, that villagers dislodged "local party agents of the Great Leap" sometimes even "twenty or thirty years after the famine" (Thaxton 2008: 293).

4 It is indicated by different character combinations that both sound like *kumai.* (In addition to 哭脉it is also sometimes indicated by 苦荬, but in either case the character for "bitter" remains in both). The English name for this leafy green vegetable is sowthistle-leaf Ixeris and its Latin name is Ixeris sonchifolia Hance. It comes in several varieties—straight edged and uneven edged.

5 Although there is not sufficient space in this chapter to go into all the examples of how the once widely available foraged foodstuffs have been commodified today, there were several other cases of this in Meixian, a particularly famous example being *yaocai* (药材), medicinal materials made from various trees that were once widely available for anyone who chose to forage in the mountains. Now, they sell in the market for a high price and are used as the basis for soups. They are consumed not so much for use as a cure, but to maintain health and defend against illness through the strengthening of different internal organs, thus enhancing bodily functions and balancing hot and cold in the body.

6 Other *zaliang* included millet (*sumi*, 粟米), dried tofu paste (*doujiao gan*, 豆角干), dried taro (*yu gan*, 芋干), corn (*yumi*, 玉米), fava bean (*hudou*, 胡豆), peanuts, and winter wheat (*xiao mai*, 小麦), which was harvested during the collective era in early March, before the first rice crop was planted.

7 Sweet potatoes not only contain many vitamins such as vitamins A, C, E and K and riboflavin, but also phosphorus, fiber, calcium and iron. Villagers in Moonshadow Pond not only eat sweet potatoes but can also cook the sweet potato leaves.

References

Buck, John Lossing. 1937. *Land Utilization in China*. Chicago: Chicago University Press.

Chen, Yunpiao. "The altar and the table: Field studies on the dietary culture of Chaoshan inhabitants." In *Changing Chinese Foodways in Asia*. Ed. David Wu and Chee-beng Tan, 19–34. Hong Kong: The Chinese University Press, 2001.

Dikötter, Frank. 2010. *Mao's Great Famine: The History of China's Most Devastating Catastrophe*. New York: Bloomsbury.

Fang Xuejia 房学嘉. 2005. *Kejia Minsu* (客家民俗. Guangzhou: Huanan Ligongdaxue Chubanshe (华南理工大学出版社).

Giddens, Anthony. 1990. *The Consequences of Modernity*. Stanford: Stanford University Press.

Holtzman, Jon. 2009. *Uncertain Tastes: Memory, Ambivalence, and the Politics of Eating in Samburu, Northern Kenya*. Berkeley: University of California Press.

Li, Lillian M. 2007. *Fighting Famine in North China*. Stanford: Stanford University Press.

Meixian Zhi. 1994. Meixian Zhi (梅县志) [Mei County Gazetteer]. Ed. Meixian Difangzhi Bianzuan Weiyuanhui 梅县地方志编纂委员会. Guangzhou: Guangdong Renmin Chubanshe.

Mintz, Sidney. 1985. *Sweetness and Power*. New York: Viking.

Ray, Krishnendu. 2001. *The Migrant's Table: Meals and Memories in Bengali-American Households*. Philadelphia: Temple University Press.

Shepler, Susan. 2011. "The real and symbolic importance of food in war: Hunger pains and big men's bellies in Sierra Leone." *Africa Today* 58(2): 43–56.

Simmonds, Frederick. 1999. *Food in China: A Cultural and Historical Inquiry*. Boca Raton, LA: CRC Press.

Song, Weijie. 2012. "Emotional topography, food memory, and bittersweet aftertaste: Liang Shiqiu and the lingering flavor of home." *Journal of Oriental Studies* 45(1/2): 89–105.

Spence, Jonathan. 1997. "'Ch'ing'." In *Food in Chinese Culture: Anthropological and Historical Perspectives*, Ed. K. C. Chang, 259–294. New Haven: Yale University Press, 1977.

Sutton, David. 2001. *Remembrance of Repasts*. New York: Berg.

Thaxton, Ralph. 2008. *Catastrophe and Contention in Rural China: Mao's Great Leap Forward Famine and the Origins of Righteous Resistance in Da Fo Village*. Cambridge: Cambridge University Press.

Wang, Zengneng (王增能). 1995. *Kejia Yinshi Wenhua* (客家饮食文化). Fujian: Jiaoyu Chubanshe (福建教育出版社), 1995.

Williams-Forson, Psyche. 2014. "'I haven't eaten if I don't have my soup and fufu': Cultural preservation through food and foodways among Ghanaian migrants in the United States." *Africa Today* 61(1): 69–87.

Yang, Jisheng. 2008. *Tombstone: The Great Chinese Famine, 1958–1962*. Ed. Edward Freidman, Guo Jian, and Stacy Mosher. Trans. Stacy Mosher and Guo Jian. New York: Farrar, Straus and Giroux.

Yang, Martin. 1945. *A Chinese Village, Taitou, Shantung Province*. New York: Columbia University Press.

Yanping Li, Yuna He, Lu Qi, Vincent W. Jaddoe, Edith J. M. Feskens, Xiaoguang Yang, Guansheng Ma, and Frank Hu. 2010. "Exposure to the Chinese famine in early life and the risk of hyperglycemia and type 2 diabetes in adulthood." *Diabetes* 59(10): 2400–2406. https://doi.org/10.2337/db10-0385.

11 "Groveling for lentils"

The culture and memory of food scarcity in occupied France

Paula Schwartz

Introduction

"We will have them groveling like dogs. ... They will trade their souls for a plateful of lentils." So speaks a cynical Nazi ideologue in the famous wartime novella *Silence of the Sea*, written and produced in the French underground by Jean Bruller, better known by his clandestine alias, Vercors. (Vercors 1942).[1] The German officer here likens the occupied French to starving animals desperately scrounging for their next meal. In the end, he maintains, hunger would edge out morality, principles, and resistance to oppression.[2]

When Vercors attributed these words to a fictitious Nazi officer in occupied Paris, he was giving voice to a sentiment the French commonly held: that the German occupiers were starving France as part of a conscious plan. Not only had the occupiers set out to dominate the French by seizing foodstuffs meant for local consumption, but many believed they also intended to *humiliate* the French, for whom food and eating were and remain integral parts of a shared national identity.

Wartime hunger revived memories of the failed harvests of the mid-to-late eighteenth century that had ended in bread riots, setting the stage for the French Revolution. Among the very oldest members of the population, it also stirred painful memories of hunger at the hands of the "hereditary enemy" to the east. During the Siege of Paris (September 1870 through January 1871) Prussian troops surrounded the capital, choking it off from the countryside and preventing foodstuffs from reaching the Parisian population. The invaders aimed to bring Paris to its knees. Their strategy worked. The French were routed in battle by the Prussians and ultimately lost the disputed province of Alsace-Lorraine. That a similar scenario of defeat and starvation be repeated less than a century later was a source of anger and shame.

Vercors wrote *Silence of the Sea* in October 1941. By 1944, countries under German control or occupation stretched from Britain's Channel Islands off the coast of France, across Europe, and deep into the western regions of the Soviet Union. They included Belgium, the Netherlands, and Luxembourg; Denmark and Norway; Greece and the northern part of Italy; Austria and other countries of Central Europe; the Balkans; and the Baltic states. At different times hunger plagued all of these countries—including Germany itself—and with varying degrees of intensity, from undernutrition and malnutrition to the most extreme cases of starvation and outright famine. In two infamous cases, the siege of Leningrad from September 1941 to January 1944, and the Dutch Hunger Winter of 1944–1945, German blockades cut off all food supplies to the local populations.

For decades, wartime hunger was a subject for economic and political historians, economists, and nutritionists, but the implications of food scarcity far exceeded the familiar story of dwindling supply and frustrated demand, or even the ravaged bodies of starving civilians. When scholars turned their attention to the study of everyday life, the impact of hunger on social relationships, mentalities, and grassroots politics showed that food scarcity had virtually reshaped public and private lives (Veillon 1995; Schwartz 1999; Schwartz and Veillon 2000; Mouré and Schwartz 2007). Hunger and the pursuit of food were not merely the background against which other stories played out but rather they became a story of their own. A rich trove of memoirs and testimony of the war years, together with the opening of police archives and those of the Ministry of Food Supply, revealed how wartime shortages transformed social relationships, fostered new economies of exchange, informed a lively popular culture, and sparked political protest.

Food scarcity: causes and consequences

German occupiers confiscated food from all over Europe to feed the beleaguered Wehrmacht on the Eastern front, and later even German civilians. Of all the occupied countries, France furnished Germany with the largest share of its industrial and agricultural production. The armistice signed in 1940 by the Vichy regime, or the French collaboration government, provided for the surrender of foodstuffs and other supplies. German occupation authorities had the power to requisition food in addition to other materials and equipment. As the war dragged on and the tide began to turn against them, German occupiers exacted ever more from France.

The Germans wielded food as a weapon. Control of food supply was used to subjugate some, promote and reward others. Regulating access to food was a powerful means of social control of occupied populations—up to a point—and provisioning the Fatherland was crucial in sustaining public support for the war at home. Access to food was also critical to the overarching project of exploiting prisoners for the war effort. For the millions of Europeans who were on the receiving end of that weapon, hunger was the single most-defining aspect of life under German occupation (Tönsmeyer 2018).

Occupying troops could help themselves to food directly from French farms, warehouses, and stores. By 1941, however, there was less to plunder. Agricultural production had already been curtailed by German requisitions of livestock, work animals, and fertilizer (used for explosives). Another consequence of the French defeat in 1940 was the absence of 1.5 million working men who were prisoners of war in Germany, which resulted in a severe shortage of agricultural labor and left farm and factory work to women. Distribution was a disaster: First, there was little to distribute because peasants held back production instead of sending it to legal markets. Second, transport vehicles, especially trains, had been commandeered for the German war effort. Intra-European trade and food imports from the overseas colonies to the metropole had been halted. Add to that the damage to infrastructure, especially by the end of the war, and what food was produced was simply not moving. This had a disastrous effect on cities, especially those in regions where agricultural production was less diversified (Mouré and Schwartz 2007).

Many French blamed the food shortages on German requisitions, at least initially, but scarcity had other causes. Some, like poor harvests, were the result of "natural" causes; most, like failed rationing and distribution systems, were caused by humans. In the division of administrative labor between French and Germans, the French authorities had control of provisioning the cities and towns. The civilian rationing system put in place in the summer and fall of 1940 was never sufficient to meet the needs of the population. The last rationing restrictions were not lifted until well after the war, in 1949. Even at their most generous, rations of dairy products, fats, meat, and bread never provided the minimal caloric and nutritional requirements for any category of the population. Oftentimes supplies ran out before rations could even be fulfilled. It was simply not possible to survive without augmenting one's food needs from other sources. Food producers hid or hoarded goods for themselves or sold directly to consumers. An underground traffic in food and other commodities, the black market, emerged to fill the gap (Grenard 2014, 2008).

In short, the French food system was profoundly altered and, with it, the relationships between city dwellers and farmers, food-producing and non-producing regions, shopkeepers and patrons, and among shoppers themselves. Food scarcity exacerbated tensions between haves and have-nots, and sometimes inverted their roles. Merchants, not clients, were king. Farmers now had power and prestige over townspeople. In another inversion, the unemployed benefited when working people were disadvantaged. Workers, especially single men and women, had less time to stand in line for hours in front of shops, many of which were often closed or were empty by quitting time, whereas the unemployed were available to rent themselves out as proxies and stand in line for others. Yet while some social hierarchies were inverted, preexisting inequalities became more deeply inscribed between the prosperous, who could afford to supply themselves on the black market, and the poor, who were left to rely on rations, oftentimes unfulfilled when stocks ran out.

Hunger had devastating effects on health, well-being, and morale, and it threatened the very survival of the most vulnerable members of French society. Morbidity rates skyrocketed. Mortality in all categories of the population rose; infant mortality rates were especially alarming and were comparable to those of developing countries today. There was a rise in stillbirths and childhood diseases like rickets. Children of the war years had slower growth rates and were smaller and shorter on average than the adults of the preceding generation (Veillon 1995, Alary 2006, 2009). Hunger also meant poor performance in school and less productivity on the job.

For many, if not most, French people, the fact of hunger came to define day-to-day existence. Marie-Louise Claudé, a prisoner-of-war wife who hid a German soldier after his desertion from the Wehrmacht, summarized her experience of the war years in two words: hunger and fear (Claudé 1987). Hunger and fear, fear of hunger: they dogged everyone, though not equally. Even those spared a relative freedom from total deprivation—such as food-producing farmers and anyone who could claim to know one, or the wealthy and privileged, who could buy on the black market—were consumed by preoccupation with food and eating.

Since daily intake had to be supplemented from sources other than official ones, institutions were particularly hard put to feed the populations in their charge. Sickness and death from starvation were disproportionately high in hospitals, boarding schools, insane asylums, orphanages, and internment and holding camps (von Beultzingsloewen 2005). Jews had to observe earlier curfews than others and could not be served in many shops. French and foreign Jews and other "enemies of the Reich" who were deported to concentration camps from spring 1942 until fall 1944 had even less to eat. This was part of a concerted plan to accelerate their demise after appropriating their capital and extracting their labor power.

Among those most severely affected by the inaccessibility of food were resisters who lived fully underground using false names and with forged identification papers, and rural partisans hiding in the woods. Accessing their rations under their own names was out of the question; these individuals and groups were wholly dependent on support from the surrounding population. Aiding and abetting a resister, feeding the partisans, or giving food and lodging to Jews in hiding were punishable by death. Underground networks specialized in the supply of false ration cards and tickets. Rural partisans raided town halls and local offices, where they stole equipment and ration books. Provisioning resisters was an integral part of the organized underground movement, one that often fell to women (Schwartz 1987).

Living with hunger

Alterations in the food system profoundly affected the culture of everyday life. Newspapers were replete with articles about food scarcity and food distribution, advertisements for ersatz products, and humorous drawings and stories. Advice columns were devoted to new, innovative recipes, strategies for making do and getting by, and substitutes for common ingredients. Scarcity affected relationships between town and country, merchants and clients, and among shoppers themselves.

Hunger may have been painful, but it was not off limits to humorists. A cartoon with the caption "What Young Ladies Dream Of" features two women leaning over a perfume counter where bottles are labelled "Whiff of Steak and Potatoes" and "Scent of Roast Leg of Lamb with Garlic" (AN, Collection Florien de Rochesnard, n.d.). The penury of food, textiles, and raw and finished materials of every sort was a common theme in jokes, songs, skits, and stories (Mouré and Schwartz 2007).

Ephemera produced in this period reveals the ubiquity of food-related concerns. Correspondents sent humorous, pre-printed postcards on which they checked the description that best suited the state of their food supply: nothing, next-to-nothing, less-than-nothing. Wedding invitations featured drawings of empty banquet tables. *Billets doux* or pre-printed love letters took the shape of ration tickets (AN, Collection Florien de Rochesnard, n.d.). An ashtray depicting the much-despised staple of the period, the infamous rutabaga, bears the inscription, "When you don't have what you like, you have to like what you have" (BDIC n.d.).

Among the many adaptations to the food crisis were new modes of cooking. Fuel was as scarce as food. Home economists urged French housewives to conserve energy by adopting a makeshift cooking method for dishes that required long cooking times, like soups, stews, and even roasts. The so-called Norwegian cooking pot (*marmite norvégienne*) was a graduated series of boxes, one inside the other, lined with newspaper or straw to retain heat. After placing a cooking implement on the stove long enough to thoroughly heat the contents, it would be transferred to the insulated series of boxes and covered. Several hours later, the one-pot meal would be ready (Chouard 1941). This method was also enthusiastically endorsed by M. F. K. Fisher, who advised American women to conserve cooking fuel in this way during wartime. She called the receptacle by a different name, the "haybox" (Fisher [1942] 1988).

People took to raising their own "livestock" for consumption. Even city dwellers raised rabbits and chickens on apartment balconies. A veterinarian published a guide called *Meat Without Tickets*:

> This little book has no scientific pretention, and I will have attained my objective if it helps ordinary people replace the *biftecks* that are temporarily absent from our menus with meat that they can produce themselves under the most economical conditions.
>
> (Pérol 1944)

Other "cookbooks" included recipes for such dishes as "Beef stew without beef," "pâté without meat," and "mackerel or herrings in white wine (without white wine)" (Deaubret 1940).

Food scarcity also transformed the urban landscape. Lines of anxious shoppers stretching along sidewalks and spilling into streets replaced familiar scenes of bustling stores and markets. Storefront displays, once lavish, now featured empty bottles and cans, artfully arranged as mere decoration. In photographs of the period, the signs in store windows warned customers,

"Everything is fake" (*Tout est factice*) (Vernay 1982; Ory 1994). It was an effort to avoid raising false hopes and no less, perhaps, to ward off suspicion that shopkeepers were withholding inventories.

Posters placarded in subways, buses, and kiosks exhorted consumers to put every part of every commodity to good use: conserve eggs by adding pieces of bread to your omelette; save potato peels and chestnut shells for fuel (Musée de la Résistance, Limoges, France). Some of these posters were produced by charitable organizations like the Red Cross or the *Secours rouge*; others by public authorities such as the Ministry of Food Provisioning or the Ministry of Agriculture. Advice of this sort was a feeble, even clueless, attempt by authorities to address the concerns of the population. Were they oblivious to the gravity of the situation, or merely unwilling or unable to respond in a more helpful manner?

A poster depicting a mother serving her children at table in a cozy home promoted the healthy attributes of a wartime diet, making a virtue of necessity:

> Let's eat less bread and more potatoes. Let's eat less fat; it would help digestion. Let's eat less sugar. Sugar is superfluous. Let's drink more water, fewer fermented drinks, and no alcohol. Eating too much meat causes stomach flu and appendicitis. Many people would be better off if they ate and drank less.
>
> ("Le Pain dans la guerre" n.d.)

Such admonitions drew wry comments from observers who noted that it was easy enough to heed advice of this nature when there was no bread, sugar, meat, or alcohol to avoid. For one privileged man who was apparently impervious to the suffering of others, food scarcity was a felicitous byproduct of war and occupation that permitted him to restore his trim silhouette (Duc de Nemours, quoted in Mouré and Schwartz 2007).

In an economy in which grooming and managing one's contacts were as crucial as money, shoppers established privileged relationships with providers. Wealth was no longer measured solely in francs; social capital was vital to the procuring of foodstuffs. Shoppers, mostly women, curried favor with food providers, whether farmers or merchants, and managed those connections carefully. In one wartime account of life under the occupation, a woman asks not whether her new daughter-in-law knows how to cook, but rather whether she will have the social skills to cultivate and maintain sellers, especially those who trucked with the black market (Toklas 1954). While some could afford to pay a price that exceeded the legal limit, savvy customers bartered for food by offering services in exchange.

Food scarcity affected other "economies" as well. Economic and social capital of the sort mentioned above enhanced human capital. In the classified section of a newspaper, a man advertised for a "lady, 35–40 years old, pretty, and gentle." His self-described attributes included being "from an old family," "pure French" (code for "not Jewish or foreign"), and most telling for our purposes here, a "*well-provisioned* gourmet" (quoted by Le François 1942, italics added). Alas, we do not know whether his fetching proposition attracted any qualified female readers.

Ration tickets had exchange value in venues other than shops and markets. Restaurant customers paid for their meals with ration cards in addition to cash. Ration tickets took the place of hostess gifts: instead of offering a bottle of wine to the host or hostess—a luxury which for most was out of the question—dinner guests were invited to bring ration tickets to help defray the "cost" of the meal. The following anecdote from a French reporter was broadcast by the BBC from London on the radio show *Les Français parlent aux Français* (The French speak to the French):

> I got off the metro and exited the station with my two heavy suitcases. I called to a street sweeper who was pushing a wheelbarrow and asked if he would carry our suitcases to our hotel a hundred meters away. He accepted. I offered him 10 francs as a tip. I was expecting him to say thank-you when he said, "Would you have a ration ticket for bread instead?"
>
> ("Voyages en France," 28 August 1942)

This particular broadcast was primarily intended for the French expatriate audience in London. The dire living conditions that prevailed across the Channel qualified as news to the world outside. Within France, the collaborationist Vichy regime exercised total control over the media. Occupied inhabitants counted on the BBC as their only reliable source of information, although listening to BBC broadcasts was strictly forbidden and punishable by arrest.

Public perceptions

As early as December 1940, police patrolling the rue de Buci, a market street in central Paris, predicted that anger against the German occupiers seemed destined to escalate.

> The nature and violence of the outburst will be all the more intense because the crowd maintains that they are being *bullied on purpose* by the occupying authorities, who, according to the crowd, will continue to diminish the food supply.
>
> (APPP, report "files d'attente" 22 December 1940)

The telling language of this police report reveals that shoppers resented the Germans for taking not only their food but also their dignity. Such notions lent themselves to extravagant claims in the underground press. German soldiers were said to grease their tanks with Normandy butter. The German military commander of occupied France, Otto von Stülpnagel, was pilloried for behavior imputed to his mistress, who was alleged to bathe daily in fifty liters of milk withheld from French children (Archives Tasca, box 157, *Les Lettres françaises*, December 5, 1941).

Public displays of consumption by occupying forces fanned the flames of popular resentment. In his diary of the war years, German officer Ernst Jünger commented on the gulf between hungry Parisians and the wealthy patrons of one of Paris's most celebrated restaurants, the Tour d'Argent, located then as now in a three-story building overlooking the Seine. He described a scene in the dining room, which he likened to a flying ship (*aéronef*) soaring above the city. The restaurant was a spatial expression of social relations, with diners literally *on top*.

> As they consume the sole and the famous ducks [that are the specialties of the restaurant], the diners at the Tour d'Argent seem like gargoyles, peering down below with diabolical satisfaction at the gray ocean of roofs where the hungry people live, eking out a meager existence. *In such times, eating, eating well and eating a lot, give one a sense of power*
>
> (Jünger 1942 [2008])

The Tour d'Argent and restaurants of its class were the near-exclusive preserve of the German elite. High-level officers and administrators, together with highly privileged French people, enjoyed lavish meals in what resembled a prewar world without restrictions. This did not go unnoticed by the populace. Resentments were patent. Jünger also noted:

> Lines growing longer in front of offices open to the public and in front of shops. When I pass by in uniform, I see furtive glances marked by the deepest aversion, combined with the wish to kill. One sees on these faces what a joy it would be if we [the German occupiers] disappeared into thin air and vanished like a dream.
>
> (Jünger [1943] 1980)

That food was seen as a weapon in the war against the French is also demonstrated by this report of an incident in Lyon. In her diary entry for February 10, 1941, Léonie Villard wrote,

> The cost of living has more than trebled in six months. … The other day, a German officer seated in a tram beside Lili tried to speak to her. As she would not listen, he began to say, as if to himself, in very good French: "What a pity this pretty girl and so many others will soon be starving to death." The Stuttgart radio recently boasted that the French would be driven to grubbing for food in dustbins. They think this will frighten us into submission.
>
> (Villard 1940–1944)

Food scarcity was far from the only humiliation inflicted on the defeated and occupied nation. But in a country where food is a privileged vector of national identity, it carried added symbolic weight.

A frequent trope in autobiographically inspired fiction was that of French people becoming reduced to animals. In 1943, the communist writer and resister Edith Thomas attributed these words to the hero of her short story, "Le Professeur et les moules" (The professor and the mussels):

> A long line was already waiting patiently, a long line of women bundled up against the cold, with scarves tied under their chins. They were glued to the closed iron gate of the fish shop like ants on the cadaver of an earthworm. What more are we than ants searching for food whose ant-hill has been trampled by a boot? … As it always came back to worries about fuel and food, what had we become if not animals? Human beings come later.
>
> (Thomas 1943)

Reducing people to animals was an assault on civilization itself. "Animals stuff themselves, men eat, only discerning men know how to eat," said the famous food writer Brillat-Savarin in his precepts on the appreciation of food and the art of the table (Brillat-Savarin [1826] 2005). Dogs, ants, animals: food scarcity would make French eaters into lowly beasts scavenging for their next meal. It was a long, hard fall for a nation that prided itself on the quality of its produce, its artisanal food-making traditions, and its cuisine.

The popular perception among the French—that the enemy was out to starve the people—is supported by historical research into German war aims. In her masterful study *The Taste of War*, Lizzie Collingham shows that German fears of food insecurity were connected to all aspects of the conflict and at every stage. They were the impetus for Hunger Plan, a project to eliminate "useless eaters"—the entire populations of the occupied areas—and secure food to feed the master race (Collingham 2011). German occupiers were well aware that local populations would suffer hunger, if not outright starvation (Gerhard 2015). In some cases, forced starvation was an intentional part of the overarching plan to subdue and conquer; in others, it was the fortuitous byproduct of war, pillage, and destruction wrought by the German armies and exacerbated by the inadequacies of national authorities. In the French case, there is evidence for both.

From restlessness to protest

As people spent considerable time and energy searching for something to eat, the food line became a new locus of sociability. Shoppers waited for hours, and while waiting they talked. In keeping with long-standing tradition and the gendered division of labor, the food line was also a distinctly women's space (Schwartz 1999, 2020). The lines that formed in front of shops became a privileged source of information for the police, who monitored markets and crowds in an effort to take the temperature of the population. It took little time for French authorities to recognize that queues, tightly packed like powder kegs, could be readily combusted by the spark of frustrations and resentments great and small. As a result, uniformed policemen were routinely assigned to areas of urban affluence, in particular markets and market streets.

None were more alert to the political import of social interaction in the marketplace than the agents who were charged with patrolling market crowds, reporting on them, and transmitting comments overheard in lines. When possible, extra police protection was provided for shopkeepers who felt vulnerable or threatened at the prospect of having to manage hungry crowds. The observations of regular police patrols, both uniformed officers and plainclothesmen, were meticulously recorded; their field notes became the basis for extensive, regular reporting on food supply problems, public disturbances, and the evolution of public opinion.

From the first months of the Occupation through the post-Liberation period, police viewed markets, shops, and food lines as potentially, even inevitably, explosive urban spaces. Their worries were not unfounded. Incidents were frequent in Paris proper and in the surrounding areas. As early as December 1940, the police commissioner of the sixth arrondissement reported tensions on the rue de Buci. Even before formal reports of market place disturbances had become routine, the local commissioner alerted central headquarters to the menacing "physiognomy of the Buci market":

> At 5:00 a.m. an enormous crowd began gathering near the various shops. By 9:00 a.m. streets and sidewalks were overrun. Almost all butcher shops were closed because they had nothing to sell.
>
> In a space of 300 meters at most, it took twenty policemen to control and handle the crowd, which was more than agitated on account of the cold temperatures and the dwindling food on display.
>
> By way of example, the sum total of available merchandise at the large Eco market amounted to only 150 rabbits in halved portions. Although they had been apprised of the situation, a line of approximately two thousand people formed nearby. … After the market closed, I gathered together the police officers and plainclothesmen who had been on duty. Our unanimous opinion: if this situation continues, hunger and cold will lead to unrest. … I have no reservation in saying that I consider the situation very serious, based on what I saw this morning at the Buci market alone.
>
> (APPP report, 1940)

Conditions had hardly improved two years later, when another police agent submitted this report:

> Housewives note with anxiety that even at the height of the season, grocery shelves are more bare than ever. Worry and anger, even when they are not manifested in a violent way, are developing everywhere, even in the most level-headed people. ... The situation is extremely critical.
> (*Bulletin hebdomadaire* no. 55, 1942)

Tensions were such that uniformed police who patrolled an open-air market in Paris were even afraid of reprimanding repeat offenders such as restauranteurs who illegally provisioned themselves, or customers who cut in line. "All the agents have asked to be replaced by plainclothesmen, which proves how afraid they are of confronting sellers or customers whose behavior is reprehensible." His assessment continues:

> We cannot emphasize enough the seriousness of this situation, which is first and foremost a *question of public order*. One of these days there will be an actual riot and it will be too late to act. ... We repeat, in the interest of public order and the proper conduct of Paris, forceful measures must be taken as quickly as possible. If not, an incident could degenerate into a fight, the outcome of which is unpredictable.
> ("Enquête sur la marché de l'Alma" 1941)

The political climate was "pre-revolutionary," (*Bulletin hebdomadaire*, July 1942); the capital and surrounding departments seethed with "dangerous social ferment" (*Bulletin hebdomadaire*, September 1942). From the early days of the German occupation until well into 1944, French prefects, mayors, and police contemplated the hungry crowd with fear and foreboding.

If the German occupiers were vilified for their cynical use of food to humiliate the occupied, the everyday face of the food crisis was French. It was the French who had accepted the terms of the armistice giving Germans the right to requisition food, and it was they who were responsible for the management of existing resources. It was French civil servants who doled out the monthly ration tickets. Nor did French "civilians" escape public wrath—merchants accused of selling to the highest bidder, shoppers who cut in line or exploited their "priority" status, black marketeers who gave first pick to the Germans—they, too, were targets of popular grievances.

Police fears were justified and ultimately confirmed. Hunger indeed sparked protest, both spontaneous and organized. There were hundreds of food-related incidents in France, engaging anywhere from a small handful of protesters to several hundred. Like the market women who marched to Versailles to demand bread in October 1789, delegations of women marched to town halls, prefects' offices, warehouses, and food-distribution centers, where they pressed authorities to release stockpiled foodstuffs. They signed petitions and presented them to mayors and prefects. Such activity was usually peaceful, if strident. On occasion protesters' demands were met with limited, short-lived, but nonetheless satisfying success: a one-time release of potatoes, or rice, or lentils.

Women also demonstrated for food on market streets and marketplaces. There were no formal processions of women carrying banners and shouting slogans—what passed for a "demonstration" at a time when public protest was outlawed could be a furtive assembly of women lasting only a few minutes. On Mother's Day 1942, at the intersection of the rue de Buci and the rue de Seine in central Paris, a group of women stormed a shop, seized the item of the day—cans of sardines packed in oil—and threw them to the crowd. The ringleader and organizer of the operation, Madeleine Marzin, told shoppers, "Help yourselves!"; "today no tickets are necessary!" Some called it a riot; others called it looting. For the principals, it was a *demonstration,* one of two emblematic women's food protests in occupied Paris that ended in drama and arrests (Schwartz 2020).

In the capital and the surrounding area, most if not all food protests were organized by women's groups affiliated with the underground French Communist Party. Because protesters were vulnerable to arrest, hidden "protection" teams comprising partisan men, both armed and unarmed, were sometimes posted along side streets to defend them. In the summer of 1942, the rue de Buci and the rue Daguerre were the sites of women-led demonstrations for food that ended in the deaths of police and, in the case of the rue Daguerre, two onlookers. Following both incidents, scores of protesters were arrested and imprisoned; some spent the duration of the war in French prisons or the German concentration camps of Ravensbrück, Mauthausen, and Buchenwald. Those who had perpetrated the violence, or were said to have perpetrated it, faced a worse fate: execution. Some men perished by the French guillotine, others by German firing squads. The deaths of policemen changed everything: what had begun as food demonstrations, albeit illegal, were treated by authorities as *crimes against the state* and carried the death penalty (Schwartz 1999, 2007, 2020).

The demonstrations on the rue de Buci (May 31) and on the rue Daguerre (August 1) were extreme cases. What came to be known as the "Buci affair" spiraled into weeks and months of investigations and arrests. Such food protests were not to be repeated, at least not in the same form. But women continued to express their grievances in markets, town halls, and prefects' offices. Their protests are recorded in police files, monthly prefects' reports, and in the prolific underground press that was published and distributed illicitly by the resistance.

In areas of France that were not occupied by the Germans until November 1942, conditions were different. Public protest was less dangerous though nonetheless illegal. Demonstrations for food took place in Nice, Marseille, and other cities and towns, both before and after the Germans had established a full-time presence (Guillon 2003). In the working-class towns of the "forbidden zone" of northern France, wartime food protests belonged to a local culture of contestation (Taylor 2000). Women also organized in Dijon and surrounding areas to press for better rations. Some of these activities were orchestrated by the underground French Communist Party and communist women's groups, others by unaffiliated groups of women and sometimes men.

Wartime hunger and food protests were not unique to France. The last months of the war did not bring immediate relief from hunger, but they did loosen the strictures that had inhibited public displays under other repressive regimes throughout Europe. Such incidents were of concern to the Allies and duly reported in the American press. On December 7, 1944, housewives in the "poorer districts of Rome" staged a protest against black market prices. The reporter concludes that "hunger because of scarcity must perforce be borne, but hunger because of profiteering provides the elements for a popular uprising and this was what happened today" (*New York Times* 1944a). An October 4, 1944 dispatch from Switzerland relayed news of food rioting in Vienna following an announcement that food rations would be reduced by 10 percent (*New York Times* 1944b).

If German planners thought that winning the war was contingent on *starving* occupied populations, an anonymous American journalist told readers that an Allied victory could be assured by *feeding* them:

> Everywhere that the Axis rule has spread in Europe it has brought a scarcity of food. ... Deprivation of food has been an Axis weapon, shrewdly and remorselessly used. ... Wherever the Allied troops move in, one result will be certain—the civilian population will get enough to eat. ... Food in the liberated regions of Italy, and elsewhere, will save lives in more ways than one. [Allied] soldiers distributing flour in Palermo are winning the war just as much as soldiers facing machine-gun fire. The more the news gets around that honorable surrender means food the less machine-gun fire the soldiers will have to face. At such a price who will say that food, even at the cost of rationing at home, isn't cheap?
>
> (*New York Times* 1943)

According to this reporter, food provisioning would spare European lives and, by extension, the lives of American soldiers sent to protect them. Once again food is seen as vital to winning the war, be it wielded as a sword or proffered as an olive branch.

Minds and bodies, individual and collective

Some of the social, cultural, and political ramifications of food scarcity, however profound, did not outlast the war and occupation of France, yet the *memory* of food deprivation persisted well beyond the end of rationing in 1949. Food scarcity left an indelible imprint on the war generation. It shaped them, and their stories of hunger at family gatherings and dinner tables all over France shaped their offspring. Children of the 1950s and beyond grew up in households where parents and grandparents were living with the small but telling reflexes that wartime economy had imposed—a visceral avoidance of Jerusalem artichokes and rutabagas; the inability to part with random bits of string, stale crusts of bread, or leftover scraps; hoarding of food and supplies for the next catastrophe. Such postwar habits are trivial compared to death from starvation, yet they reveal how food insecurity shapes mentalities long after abundance has been restored. Wartime scarcity cast a long shadow.

Hunger also marked bodies—and not just those of the victims themselves. Children of mothers who were pregnant during the Dutch Hunger Winter, an extreme situation circumscribed in place and time—the northern and western areas of the Netherlands from January to May 1945—have made ideal test cases for the study of the medium- and long-term effects of famine. Longitudinal studies of the Hunger Winter cohort have shown that the body "remembers" hunger, and that memory is transmitted genetically from one generation to the next (Lumey 2007; Schulz 2010). Children born to women who were pregnant during the Dutch Hunger Winter bear the genetic markers of the food deprivation suffered by their mothers. Later in life they had higher levels of diabetes, schizophrenia, heart disease, and obesity. That is not where it stops: the same epigenetic markers of famine are found in *their* children (Kaiser 2014). Thus, although the subsequent generations grew up in a time of prosperity and abundance, their bodies were and are still "coded" for famine. No such studies exist for France, where food supply varied enormously by locality and by timeframe over a protracted, seven-year period, but the results of Dutch Hunger Winter research are suggestive of similar effects in other food-deprived populations.

It is estimated that some 20 million Europeans died of starvation and starvation-related disease during the Second World War (Collingham 2011). In Leningrad, a city of 3 million when Hitler invaded the Soviet Union in June 1941, 750,000 perished from hunger, most of them during the winter of 1941–42, when the Germans encircled the city, cutting it off from food supplies (Reid 2010). The Dutch Hunger Winter claimed 45,000 lives in the space of four months. How many of the civilian deaths in Second World War France were the direct or indirect result of food scarcity?

If the qualitative effects of wartime hunger in France are abundantly clear from oral testimony, popular culture, memoirs, and letters of the war period, quantitative measures fall far short. The total number of civilian deaths in France between 1940 and 1945, some 400,000, is itself a matter of some debate (Lagrou, cited by von Bueltzingsloewen 2005.) The number of deaths resulting from starvation and hunger-related illnesses remains equally obscure.

One of the numerous methodological problems associated with assembling such figures for France is a matter of language. The terminology used to account for death by starvation is as diverse as it is vague. "Dropsy," "heart failure," "kidney failure": officials registered the many cases of death from starvation by different names. Lack of food also lowers resistance to infectious disease, hence, the prevalence of the typical starvation scourges, typhus and dysentery, which were epidemic in internment and concentration camps all over Europe. All of these afflictions were symptoms. Hunger was the root cause.

The widespread use of euphemism to talk about starvation is striking. In French hospitals and psychiatric facilities, 40,000 patients perished from "inanition" (von Bueltzingsloewen 2005, 47). During the siege of Leningrad, when the Soviet government went to great lengths to keep hunger stories from reaching the outside, one spoke of "dystrophy," a new term for the diminished human body (Reid 2011, 210). The "dystrophics" dropping dead in the streets of Leningrad were like the so-called *musulmans* (Muslims) of the concentration camps—people so depleted they seemed barely human.

Euphemism as a way of naming is an effort to attenuate or conceal, but it leaves open the question of why. Why was death by starvation called by other names? Were caretakers, physicians, officials, or family members inhibited by a sense of propriety? Did fear, or shame, or guilt mandate discretion? Was it unthinkable and therefore unmentionable that first-world populations suffer third-world deaths? In any event, if researchers have yet to quantify hunger deaths for France, the staggering numbers above tell a story that transcends that of any one occupied country or region. In the final analysis, the real legacy of wartime hunger is not so much what it left behind as what it took away.

The use of diesel or paraffin oil in salad dressing, as French cooks were exhorted to do, is hardly an index of suffering on the scale of the famines that affected other occupied territories. Indeed, some historians now qualify the situation in France as a "selective famine," one that struck vulnerable populations while leaving others relatively unscathed (von Bueltzingsloewen 2005). Food was nonetheless the primary focus of everyday life under German occupation. The persistent gnawing of hunger infused French politics, culture, and society of the war years. It was a time when "What's for dinner?" was less an innocent question than the wistful expression of a challenge, a hope, or a dream.

Notes

1 All translations mine.
2 For a more in-depth treatment of the issues discussed in this essay, see Mouré and Schwartz (2007); Schwartz (1995, 2020); Schwartz and Veillon (2000), from which some of this material has been adapted.

References

Archives Nationales (AN). n.d. AB XIX 4030, "France 1939–1950," dossier 1. Unlabelled clipping with no indication of provenance or date. Paris, France.

Archives de la Préfecture de Police de Paris (APPP) B A/1808, Ministère du Ravitaillement [Ministry of Food Supply], report of 22 December 1940, files d'attente.) Emphasis mine.

Alary, Eric et al. 2006, 2009. *Les Français au quotidien, 1939–1949*. Paris: Éditions Perrin.

Bibliothèque de documentation internationale contemporaine (BDIC). Archives, Nanterre, France.

Brillat-Savarin, Jean-Anthelme. [1826] 2005. *Physiologie du goût.* Avec une lecture de Roland Barthes. Hermann Éditeurs: Paris.

Bulletin hebdomadaire no. 55. 10 July 1942. To the Ministère de l'Information, Bureau d'études, Section du Contrôle de l'Opinion, Box 159, Fondazione Feltrinelli, Archives of Angelo Tasca, Milan, Italy.

Bulletin hebdomadaire no. 65 (18 September 1942), to the Ministère de l'Information, Bureau d'études, Section du Contrôle de l'Opinion, Box 188, Fondazione Feltrinelli, Archives of Angelo Tasca, Milan, Italy.

"Enquête sur le marché de l'Alma" 23 June 1941. Rapport de M. de Cartassac, Contrôleur de la Répartition des Denrées, Préfecture de la Seine,1338 W, art. 218, Archives de la Ville de Paris.

Chouard, Pierre. 1941. *Conseils et recettes pour les temps difficiles: L'alimentation en temps de restrictions*. Paris: Éditions Gautier-Languereau, 17–18.

Claudé, Marie-Louise. 1987. Interview with author, Paris, France.

Collingham, Lizzie. 2011. *The Taste of War: World War II and the Battle for Food.* New York: Penguin Books.

Deaubret, Jacques. 1940. *Qu'allons-nous manger cet hiver?* Royan: Éditions Raymonde Fournier, 25–26.

Duc de Nemours, "Four Years in Occupied France," p. 31; unpublished memoir, Hoover Institution Archives. Quoted in Mouré and Schwartz, 2007, 267.

"Voyages en France" (28 August 1942), in Crémieux-Brilhac, Jean, ed., 1975. *Les Voix de la Liberté: Ici Londres, 1940–1944*, tome II, La Documentation Française, Paris, 199.

Fisher, M. F. K. [1942] 1988. *How to Cook a Wolf.* San Francisco: North Point Press, 16–17.

Gerhard, Gesine (2015) *Nazi Hunger Politics: A History of Food in the Third Reich,* London: Rowman & Littlefield.

Grenard, Fabrice. 2014, 2008. *La France du marché noir* (1940–1949). Paris: Payot.

Guillon, Jean-Marie. 2003. "Les manifestations de ménagères: protestation populaire et résistance féminine spécifique" in Mechtild Gilzmer et al, eds., *Les femmes dans la Résistance en France*. Paris: Tallandier, 107–133.

Jünger, Ernst. [1942] 2008. *Journaux de guerre, II. 1939–1948*. Paris: Gallimard, 316. Entry of 4 July 1942.

Jünger, Ernst. [1943] 1980. *Second journal parisien, Journal III, 1943–1945*. Christian Bourgeois éditeur, Paris, p. 170. Entry of 29 September 1943.

Kaiser, Jocelyn. 2014. "Mom's Environment during Pregnancy Can Affect Her Grandchildren." *Science*, July 10.

Lagrou, Pieter. 2002. "Les guerres, la mort, et le deuil : bilan chiffré de la Seconde Guerre mondiale." Brussels: Complexe. Cited by Isabelle von Bueltzingsloewen, "La difficile identification des victimes d'une famine sélective" in von Bueltzingsloewen, Isabelle, (ed.), 2005. *'Morts d'inanition': Famine et exclusions en France sous l'Occupation*. Rennes: Presses Universitaires de Rennes, 16.

Le François, Louis (pseudonym). 1942. *J'ai faim… ! Journal d'un Français en France depuis l'Armistice*. New York: Brentano's, 289.

"Le Pain dans la Guerre." n.d. Exhibit, Mairie du 6è arrondissement. Paris, 2010.

Les Lettres françaises, December 5, 1941, Milan, Fondazione Feltrinelli, Archives Angelo Tasca, box 157.

Lumey, L. H. 2007. "Cohort Profile: The Dutch Hunger Winter Families Study." *International Journal of Epidemiology* 36: 1196–1204.

Mouré, Ken. 2010 "Food Rationing and the Black Market in France (1940–1944)." *French History* 24(2): 262–282.

Mouré, Kenneth, and Paula Schwartz. 2007. "*On vit mal*: Food Shortages and Popular Culture in Occupied France, 1940–1944." *Food, Culture & Society* 10(2): 262–295.

New York Times. 1943. "Bread, Salt, and Victory." *New York Times*, July 29.

New York Times. 1944a. "60 Reported Killed in Vienna Food Riot." *New York Times*, October 4.

New York Times. 1944b. "Housewives Riot on Prices in Rome." *New York Times*, December 8.

Ory, Pascal. 1994. *Doisneau 40/44*. Paris: Éditions Hoëbeke, 18.

Pérol, Gustave. 1944. *Bibliothèque de la vie pratique. Viandes sans tickets*. Paris: Hachette.

Reid, Anna. 2011. *Leningrad: The Epic Siege of World War II, 1941–1944*. London: Bloomsbury Publishing.

Schulz, Laura C. 2010. "The Dutch Hunger Winter and the Developmental Origins of Health and Disease." *Proceedings of the National Academy of Sciences* 107(39): 16757–16758.

Schwartz, Paula. 1987. "Redefining Resistance: The Activism of Women in Wartime France," in Margaret Randolph Higonnet et al., eds., *Behind the Lines: Gender and the Two World Wars*. New Haven: Yale University Press.

Schwartz, Paula. 1999. "The Politics of Food and Gender in Occupied Paris." *Modern and Contemporary France* 7(1, February): 35–45.

Schwartz, Paula. 2020. *Today Sardines Are Not for Sale: A Street Protest in Occupied Paris*. New York: Oxford University Press.

Schwartz, Paula, and Dominique Veillon. 2000. "Ravitaillement, pouvoirs publics et protestations dans Paris occupé," *Mélanges de l'École française de Rome,* tome 112(2): 645–653.

Taylor, Lynn. 2000. *Between Resistance and Collaboration: Popular Protest in Northern France, 1940–45*. New York: St. Martin's Press.

Thomas, Edith. 1943. "Les Moules et le professeur." In *Contes d'Auxois. Transcrit du réel.* Paris: Éditions de Minuit, 10.

Toklas, Alice B. 1954. *The Alice B. Toklas Cook Book*. New York: Lyons Press, 207.

Tönsmeyer, Tatjana. 2018. "Supply Situations: National Socialist Policies of Exploitation and Economies of Shortage in Occupied Societies During World War II," in Tatjana Tönsmeyer, et al. eds, *Coping with Hunger and Shortage under German Occupation in World War II*. United States: Palgrave Macmillan. 3–23.

Veillon, Dominque. 1995. *Vivre et survivre en France, 1939–1947*. Paris: Payot.

Vercors [Jean Bruller]. 1942. *Silence de la mer*. Paris: Éditions de Minuit.

Vernay, Laurie. 1982. *Lyon des restrictions.* Grenoble: Éditions de Terre et de Mer, Éditions des 4 Seigneurs.

Villard, Léonie. 1940–1944. Untitled manuscript, p. 16. Mount Holyoke College Archives and Special Collections, South Hadley, Massachusetts.

von Bueltzingsloewen, Isabelle. ed. 2005. *'Morts d'inanition': Famine et exclusions en France sous l'Occupation*. Rennes: Presses Universitaires de Rennes.

12 Coping with food safety risks

Information sources and responses by residents in Japan in the aftermath of the Fukushima nuclear accident

Tomiko Yamaguchi and Joo-Young Jung

Introduction

The disaster caused by the Great East Japan Earthquake, a tsunami, and the subsequent explosion of nuclear power plants in Japan on March 11, 2011 was unprecedented in complexity, magnitude, and extent of damage. According to official government records, at the peak of the crisis, in 2012, approximately 170,000 people out of 2 million in Fukushima Prefecture were displaced from their homes to shelters or to friends and family.[1] Damage to the agricultural sector was also severe because the region contributed greatly to the agricultural sector of Japan; the loss was estimated at 2,400 billion yen (approximately US$24 billion) worth of agricultural land, infrastructure, farm supplies, and crops.[2] The disaster has imposed a great many challenges on the residents in the area, especially to their livelihoods. The lives of people living in the cities in the eastern part of Japan were also disrupted in multiple ways, generating social confusion and fears. For instance, when radioactive iodine was found in a water-purification facility in Tokyo, operated by the Tokyo Metropolitan Government, panic struck the residents and many literally ran to the stores and vending machines to try to stock up on "safe" mineral water (Sports Hochi 2011). The media every day showed images of empty convenience store shelves in Tokyo and people queueing to buy food, water and gasoline (e.g., Asahi Shinbun Digital 2011). These experiences revealed that the seemingly stable food supply in Japan rests on a fragile and vulnerable food system.

The extreme troubles described above were triggered by the disaster, but the fragility and vulnerability of the system are more chronic issues in Japan and are often framed in public-policy discourse as a matter of the nation's food security. Indeed, Japan's food system is fragile due to the fact that Japan's food self-sufficiency rate is low and that Japan's "food miles," the distance food is transported from production site to consumption sites, are the longest among Organization for Economic Cooperation and Development (OECD) countries (Nakata 2003). Fragility can also be explained as a result of much broader social changes, such as food globalization processes driven by the industrial and corporate agricapitalist food system (McMichael 1994), an aging farm population that is growing, shrinking farms (Masters et al. 2013), environmental constraints on producing food (Huang et al. 2011), and changing patterns of consumer choice and food consumption (Beghin 2006). All of these phenomena have been influencing Japan's food system.

It is in this context that the disaster-hit regions have long been essential to Japan's agricultural sectors. The Great East Japan Earthquake has further complicated Japan's food-security issues by adding more fragility to the system and stoking the public's fears of radiation contamination in food. The fears of exposure to radioactive nuclides in the air, water, and food tended to be intensified by the very invisibility of the threat (Kuchinskaya 2014). Adding further complexity is the fact that the disaster made people recognize that risks involve social inequalities, meaning that the wealthy in terms of income, power, and education have access to "better" information that will help them avoid risks (Beck 1992).

Against this backdrop, our chapter aims to explore the question of how lay people responded to the perceived fragility of food systems in the wake of the disaster, focusing particularly on food-consumption behavior. Taking as a departure point the idea that widespread concerns over food safety in contemporary Japan are "a logical outgrowth of historical issues about food security and health" (Jussaume et al. 2017), we can see that understanding issues of food safety is important to understanding food security issues. Based on a survey taken among residents of Fukushima and Tokyo, we focus in this chapter on consumers' behavior in two different locations, one in the area directly affected by the disaster and the other in an area that depends on food supplies from this area; we look at their reasons for engaging in certain behaviors and the types of information they used to cope with the situation. We will first detail the methods used for this study.

Research methods

The study set forth in this chapter uses data from a survey conducted in June 2014 (three years after the disaster) with residents of Fukushima and Tokyo, as well as informal interviews with residents of Fukushima. The survey measured changes in respondents' consumption behaviors after the disaster and documented the information sources they relied on to make decisions.[3] The respondents for our survey were recruited from a panel provided by a survey research firm. Two hundred responses from Fukushima (N = 100) and Tokyo (N = 100) were collected. Respondents in Fukushima and Tokyo were selected to be among the population most likely to have concerns about food safety: females between the ages of 20 and 59 who have children between the ages of 0 and 14, and who cook three or more times per week. These criteria were chosen because the literature suggests that demographic factors have a significant influence on people's response to risks (Nayga 1996): Women are more likely to be conscious of food safety than men (Dosman, Adamowicz, and Hrudey 2001; Sugimoto et al. 2014), and those who have children in the household are more likely to be concerned about food safety (Hamilton 1985; Jussaume and Judson 2010). Cooking three or more times per week was included as one of the filtering criteria based on the assumption that people who cook often are likely to make more conscious efforts to obtain safe food than those who do not cook. In addition to the 200 participants in the survey, the study includes data from interviews with a group of mothers who are actively engaged with group activities regarding food safety concerns. We include the interviews to provide richer context to understand the survey results.

Studies suggest varying assessments of the relationship between proximity and food risk concerns. Tokyo is about 250 km from the Fukushima nuclear plants. The radiation level from the Fukushima nuclear accident was much higher in Fukushima than in Tokyo. But when it comes to food safety, Tokyo was affected because of its reliance on northeastern Japan for food. Sugimoto and colleagues (2014) suggest that consumers in eastern Japan, where Tokyo is located, are significantly more concerned about food safety than residents of other parts of Japan. Another study indicates that residents of Tokyo have a greater perception of risk than residents of Miyagi Prefecture, which is adjacent to Fukushima Prefecture (Goodwin et al. 2012). These studies are based on the premise that people farther from a disaster may actually be more frightened than people who are closer, because imagining the worst produces greater fear than seeing directly how bad things are (Dolinski et al. 1987; Halpern-Felsher et al. 2001). Other studies suggest that direct experience is likely to be associated with higher risk perceptions (Weinstein, et al. 2000; Ho et al. 2008). Given the complicated nature of the relationship between proximity and food safety concerns, whether residents in Fukushima and Tokyo differ in their food-consumption behaviors is an important analytic for understanding the relationship between geographical locations and lay people's responses to worries about food quality, or food insecurity more broadly.

Review of literature

Reasons for food safety concerns

Studies suggest that concerns over food safety are shaped not only by calculable or measurable hazards—the kinds of "solid facts" that can be detected by technical means such as radioactive nuclides, but are mediated by a range of social psychological factors, such as the conviction of having personal control over the magnitude or probability of the risk and the degree of familiarity with the risk (Slovic et al. 1981). Food safety concerns are additionally influenced by social, cultural, and institutional processes (Wynne 1996). For instance, people may avoid a particular food because they might not trust the information they have obtained from the government (Korthals 2015), perhaps because they view the government as unduly influenced by certain social actors such as people in the industry who have vested interests (Todt et al. 2009). Another reason for the lack of trust is skepticism of the government's competence in handling the problems (Reiher 2017). Further, there could be a situation in which bureaucracy may prevent the government from releasing information in a timely manner. During the aftermath of the disaster, people in Japan experienced numerous incidents in which the government delayed providing much needed and urgent information, such as the fact that the reactor core had melted down. The data generated by the simulation device called the System for Prediction of Environmental Emergency Dose Information (SPEEDI), a forecast that could have provided useful information for deciding where to evacuate, was also not provided to the public (Sugawara and Juraku 2018).[4] These contributed to a growing government distrust, thus turning people to other avenues to obtain information to decide what to do with perceived problems.

Safety concerns and the use of information

When confronted with ambiguity in coping with a disaster, people feel a heightened need for information. People's dependency on the mass media increases in a disaster situation (Hirschburg, Dillman, and Ball-Rokeach 1986; Loges 1994). People also seek out information from a variety of sources such as interpersonal contacts, experts, and social institutions (Berger 1987; Harmon-Jones and Mills 1999). Indeed, people depended heavily on the mass media after the Great East Japan Earthquake (Johnston 2012; Jung 2012; Yonekura 2013).

There is some indication that the types of information sources people use are likely to influence their perceptions of food safety and their consumption behaviors. While the mass media continue to occupy an important position in shaping people's perceptions, more and more people, particularly young people, are relying on the newer media, such as the Internet and social media, to acquire food safety related information. Past studies have found that people seek out information from newer media when their uncertainty about a situation is not sufficiently resolved by traditional mass media (Dosman et al. 2001; Overbey et al. 2017). In China, the use of Weibo, a dominant national social network service (SNS), was positively related to awareness over food safety and preventive action, whereas time spent consuming television news, newspapers, magazines, and online news was not significantly correlated with awareness over food safety (Mou and Lin 2014).

In addition to the old and new media, various other kinds of information are consulted for making food-consumption choices, such as web pages, booklets, and fliers provided by governmental institutions, scientific establishments, universities, consumer groups, food retailers, and clinics (Moldovan-Johnson et al. 2014). A survey by Japan's Food Safety Commission indicates that the percentage of people who use television to obtain information about the health effects of food containing radioactive substances has declined since the earthquake, while the percentage of people who use academic and technical books, as well as information disseminated from university and research institutes, has increased since the Great East Japan Earthquake (Food Safety Commission 2014). This phenomenon suggests that in the aftermath of a disaster the public may increasingly seek technical information beyond mass media.

People's immediate social networks, such as family, friends, and co-workers, also play significant roles in influencing people's perceptions of food safety and food-consumption decisions. If individuals perceive their reference group to be highly concerned about risks related to the nuclear disaster, they are more likely to have higher risk perceptions (Goodwin et al. 2012). Further, while people relied on the mass media far more than on interpersonal contacts to understand the overall situation after the Great East Japan Earthquake, their reliance on interpersonal networks was as high as their reliance on the mass media for making behavioral decisions, such as whether to change their food-consumption patterns or to evacuate from Tokyo immediately after the earthquake (Jung 2012).

Governmental strategies

In dealing with the public's food safety concerns, the Japanese government pursued various strategies because it saw the public's negative attitudes toward food as a hindrance to reconstruction efforts (Chino 2014). Among them is the strategy built into the government's risk-communication programs, such as the ones implemented by the Consumer Affairs Agency. The agency has carried out more than 150 risk-communication programs a year since the disaster in various prefectures in northeastern Japan.[5] In fact these programs became a vehicle for spreading "food safety culture" (Baur et al. 2017). As Renn (1992) points out, one of the purposes of risk communication is to persuade people to change their attitudes or behavior. With that in mind the agency attempted to disseminate the message that the food safety problems are under control by emphasizing the low level of radioactive nuclides in food and the existence of strict regulation standards. The present standards established in Japan for radioactive cesium in food and beverages state that the limit for general food products such as fruits, vegetables, rice, seafood, and meat is 100 Bq/kg. The limit for milk, baby food, and infant formula is 50 Bq/kg. The limit is 10 Bq/kg for drinking water and tea leaves.[6] Underlying the thinking in these risk-communication programs is the assumption that the lay public suffers from a "knowledge deficit" (Irwin and Wynne 1996). Thus, programs are designed in such a way that the public, such as residents of Fukushima and consumers in cities, are able to understand the scientific interpretation of food safety.

However, the government-sponsored programs have not had the full effect that policymakers wished for. Skepticism persisted long after the disaster, as evidenced by the social phenomena that food items from Fukushima remain unsold at the stores (Yamaguchi 2016). These phenomena suggest that changes of attitudes and behavior toward food are not as simple as the risk-communication programs suggested, because the content of these programs is embedded in wider social and political contexts: communication of risks is not about hazardous substances per se, but involves personal and social responsibilities, accountability and trust in government and in food systems (Cvetkovich and Löfstedt 1999).

Findings and discussion

The complexity of changing food-consumption behaviors

The first notable finding from our survey is that concerns about radiation contamination in food have led a large proportion of respondents to change their food purchasing behavior. Overall, 65.7% of the respondents—including respondents in both Fukushima and Tokyo—reported that they had changed food-consumption behaviors (changed: 27.8%, somewhat changed: 37.9%), whereas 34.4% said they had not (did not change: 7.1%, did not change much: 27.3%). When this data is disaggregated, 73% of Fukushima respondents changed their food-consumption behaviors (M = 3.30) while 53% of Tokyo residents did (M = 2.70). Controlling for income, education, and age, the result of the analysis of covariance indicates that the mean difference between Fukushima and Tokyo respondents in the likelihood of changing food-consumption behaviors was statistically significant ($f [165] = 6.75$, $p < .05$).

Those who responded that they had changed their food-consumption behaviors after the disaster disclosed the following salient ways in which they had changed:

- "I confirm the origin of food items when shopping for groceries." (67.7%)
- "I mainly buy food items sourced from western Japan." (30.8%)
- "I inquire whether grocery shops or distributors conduct radiation screening on the products they carry." (26.9%)
- "I buy groceries to stock for emergency situations." (25.4%)

These results indicate that sourcing food items from prefectures that are farther away from the nuclear power plants has become the prevailing and most important strategy people use to find food items they consider safe. In our open-ended inquiries, respondents also indicated that the origin of food items became an important factor for selecting safe food. When they could not confirm the origin of food items, some of the respondents avoided purchasing those food items entirely. Indeed, in one interview a woman who lives in Fukushima and has children mentioned that vegetables grown in the local area often remain unsold in the supermarket when products from

Table 12.1 Reasons for changing food-consumption behaviors

Reasons for changing food-consumption behaviors	*% of respondents (N=131)*
I am worried about radiation contamination in general.	71.5%
I am concerned about the sufficiency of regulation and monitoring of radiation contamination in food items.	39.2%
I am uncertain about the long-term effects of radiation exposure.	38.5%
News and other information sources indicate that the radiation contamination still continues.	31.5%
I am suspicious about the scientific basis for regulations on radiation.	24.6%
I am suspicious whether business operators sufficiently comply with law and regulations.	15.4%

other prefectures are available, even if the items sourced from Fukushima are offered at a lower price.

The survey also asked the main reasons for changing or not changing food-consumption behaviors. Table 12.1 shows the percentage of respondents who chose each reason for changing food-consumption habits. The top reasons for changing food-consumption patterns were: "I am worried about radiation contamination in general" (71.5%); "I am concerned about the sufficiency of regulation and monitoring of radiation contamination in food items" (39.2%); and "I am uncertain about the long-term effects of radiation exposure" (38.5%).

These results indicate that the worries driving such behavioral changes are not allayed either by governmental regulations or by campaigns to persuade people to trust the testing regimens. Instead, people look to food labels or use other mechanisms, such as consulting their personal networks or online food-purchasing sites to source food items that are perceived as safer. Further, people's perceptions of the danger or safety of food tended to be based more on the origin of food items than on the government's monitoring system.

The transition from worrying to changing food purchasing behavior (and thus sourcing food items from prefectures farther away from the accident site) is not as straightforward as one might think from the survey results, because people sometimes face difficult moral dilemmas arising from the very social networks to which they turn for information. A mother with children residing in Fukushima mentioned that she is torn between the idea of supporting farmers and fisherman in Fukushima and concern about the potential long-term health effects, especially for her children, of eating what may turn out to be radiation-contaminated food. Sharing a similar sentiment, another woman in Fukushima described how her relations with her mother-in-law turned sour when she began to avoid buying vegetables and rice from the area. Her mother-in-law told her she was too sensitive. Another woman shared her guilt about feeling uncomfortable using vegetables given to her by neighbors who had grown the crops on their farms.

The efforts by local governments and retailers to encourage consumption of Fukushima produce—campaigns with names such as "Cheer up Fukushima!" and "Eat Locally"—have attempted to allay people's concerns about the safety of local products and at the same time encourage a sense of loyalty (and even duty) to support Fukushima farmers and fishermen. We can see from our survey results that such efforts, along with other social pressures, can cause conflicted feelings; many people faced moral dilemmas as individuals and as members of society.

Information sources used for deciding food purchases

In order to cope with uncertainty in food safety, people consulted various types of information when making decisions about food consumption. Among the 19 information sources listed in the survey, respondents were most likely to be influenced by television/radio (M = 3.05, SD = 1.45) followed by husband/partner (M = 3.00, SD = 1.57) and newspapers/magazines (M = 2.96, SD = 1.52) on a five-point scale ranging from 'low influence' to 'high influence.' Information from the local government, their mothers, university/research institute websites, and governmental websites were the next most influential sources (Figure 12.1). The results indicate that media, experts, and interpersonal sources play comparable roles in influencing people's decision-making processes with respect to food consumption.

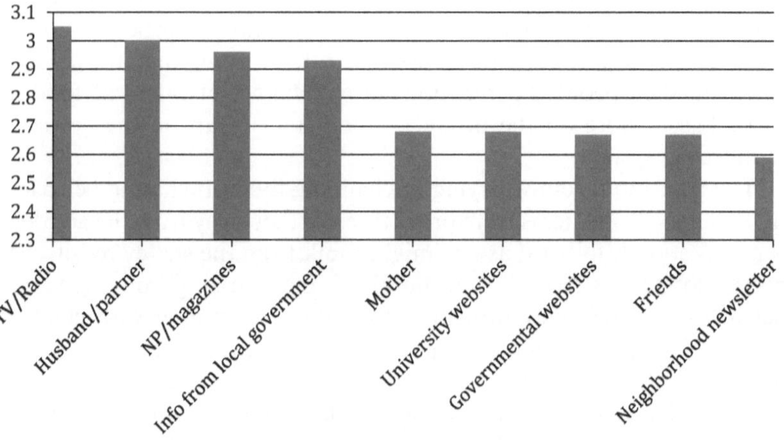

Figure 12.1 Influential information sources for making decisions on food consumption. Note: Range 0 to 5 (0 = no influence, 1 = low influence, 5 = high influence).

Table 12.2 The effect of food-consumption change on the overall reliance on information sources (Analysis of Covariance)

	df	*F*	η	*p*
Age	1	7.708**	.045	.006
Income	1	2.121	.013	.147
Education	1	1.525	.009	.219
Residency	1	.143	.001	.706
Change in food consumption	1	3.983*	.024	.048

Note: *: $p < .05$, **: $p < .01$, R squared = .103.

The survey also inquired whether there is a relationship between the overall reliance on information sources and food-consumption behaviors. Analysis of covariance was conducted to test the difference in the overall degree of reliance on information sources between those who changed and those who did not change food-consumption behaviors—controlling for education, income, age, and residency (Fukushima or Tokyo). The results (Table 12.2) indicate that those who changed food-consumption behaviors had a significantly higher degree of reliance on information sources (M = 2.70) than those who did not change food-consumption behaviors (M = 2.37, $F [1, 163] = 3.98$, $p < .05$).

Another question sought to determine whether there is a relationship between types of information sources people rely on and food-consumption behaviors. Analysis of covariance was conducted to test for differences between those who changed and those who did not change food-consumption behaviors, in terms of reliance on each of the 19 types of information sources. Education, income, age, and residency were controlled. Those who changed food-consumption behaviors were significantly more likely to be influenced by the following four information sources than those who did not change food consumption: Social Networking Service (SNS)/blogs (changed: M = 2.43, not changed: M = 1.72; $F[1, 164] = 10.97$, $p < .01$); university/research institute websites (changed: M = 2.93, not changed: M = 2.29; $F[1, 164] = 6.88$, $p < .05$); newspaper/magazine (changed: M = 3.16, not changed: M = 2.66; $F[1, 164] = 4.72$, $p < .05$); and one's mother (changed: M = 2.83, not changed: M = 2.31; $F[1, 164] = 4.46$, $p < .05$) (Table 12.3). For other information sources, the difference was not significant. The result indicates that the most prominent difference between those who changed food-consumption behaviors and those who did not change pertained to SNS/blogs. This result is consistent with other studies conducted after the Fukushima nuclear accident. Many people whose concerns and uncertainty about food safety were not resolved by the mass media relied on SNS/blogs to actively seek out information about food safety. With specific reference to food risk and benefit communication, Rutsaert, Pieniak, Regan et al. (2014) suggest that stakeholders and experts alike view SNS as a speedy and accessible medium

Table 12.3 The effect on food-consumption change of reliance on specific information sources (Analysis of Covariance)

	SNS/Blogs		*University/ Research Institute Websites*		*Newspapers/ Magazines*		*Mother*	
	F	η	*F*	η	*F*	η	*F*	η
Age	6.737*	.039	1.623	.010	.010	.000	10.471**	.060
Income	1.590	.010	5.506*	.032	.455	.003	.579	.004
Education	.865	.005	12.520**	.071	1.146	.007	.087	.001
Residency	4.761*	.028	.058	.000	.015	.000	1.142	.007
Food-consumption change	10.970**	.063	6.884*	.040	4.715*	.028	4.460*	.026
R Squared	.159		.192		.044		.101	

Note: *: p < .05, **: p < .01.

of information. With regard to demographic factors, those who are younger were more likely to rely on SNS/blogs and their mothers than did those who are older. Tokyo residents were more likely to rely on SNS/blogs than were Fukushima residents. Those with higher income and education were significantly more likely to rely on university/research institute websites than were others (Table 12.3).

Conclusions

In this chapter, using a survey and interviews with the residents in Tokyo and Fukushima, we examined how lay people responded to concerns over food safety after the Fukushima nuclear accident. The study results indicate that most respondents are consciously monitoring the safety of food in relation to radiation contamination. About two-thirds of the respondents changed their food-consumption behaviors, and the majority of them reported having uncertain feelings about the safety of food products and the health effects of radiation. Our results indicated that, while Fukushima residents were more likely to have changed their food consumption-behaviors than Tokyo residents, many faced a moral dilemma between assuring food safety for their family and supporting the farms in their prefecture.

Our study found that many people relied on multiple sources of information to cope with food safety concerns, and that people who relied on a wider range of information sources were more likely than others to take action to avoid contaminated food. With regard to specific information sources, those who changed food-consumption behaviors were significantly more likely to be influenced by new media platforms and specialized information sources such as SNS/blogs and university websites than those who did not change food-consumption behaviors. Despite the government's efforts to communicate to the public that it is safe to consume food products from Fukushima and adjacent areas, our study results indicate that not many people relied on governmental sources to cope with concerns over food safety. These findings are interesting in light of the government's approach to risk over food safety. The government is taking a scientific and managerial approach to risks based on the assumption that scientific knowledge is a vital resource for determining the truth, running campaigns to convince people that the existing inspections and regulations are sufficient to assure food safety. There is ample evidence that the public has heard and understood these views of the experts (Sano 2014), and yet some people continue to avoid purchasing food products suspected of containing radioactive nuclides. These phenomena demonstrate that, even in facing an emergency situation when food availability and food access are unstable, people attempt to get access to food that meets their normative expectations. This shows that food-consumption behavior is deeply rooted in individuals' norms and values (Morgan et al. 2006).

Our studies indicate that consumers who changed their food-consumption patterns in response to the accident were not, as the government's position would assume, suffering from a knowledge deficit. In fact, they were more likely to have sought out information and sought it from a wider variety of sources. The government's belief that "more information" or "better information" will persuade consumers to accept foods from the Fukushima area as "safe" is a belief that fails to take the full picture into account. In order to be successful in communicating to the public that food products from the affected areas are safe, the government should more actively reach out to the public via various information sources the public relies on. Our study indicates that not only the mainstream media but also new media platforms, such as social media, are important channels to disseminate information to residents, particularly to the younger generation. Additionally, our study indicates that word of mouth (e.g., from family and friends) is an important way in which people obtain information. Efforts can also be made to disseminate information via local communities and organizations to effectively reach the public. Unless and until those who design the government's communication strategies recognize that substantive concerns over food safety are tied up with more complex societal and psychological forces and processes, the mismatch between the government's expectations and the reality on the ground will continue.

The findings of the present study have implications for the nation's food security policy. Various laws, ordinances and guidelines exist for dealing with food security issues. The one that deals with problems related to food security issues in emergency situations is the *Emergency Food Security Guidelines,* issued in September 2012,[7] which shows a plan of action and the scope of responsibilities of concerned parties such as government ministries, local government offices, various interest groups and the general public. Accompanying the guidelines is the report titled *Analysis and Evaluation of Food Supply Risks,* published in March 2019, which identifies issues and a concrete plan of action for dealing with food security issues in the food supply chains, from production, processing, and distribution to consumption. Although these documents elaborate strategies for dealing with such issues, such as the reconstruction of infrastructure for production, processing and distribution of food, as well as hazard preparedness, less emphasis is given to the importance of the government's actions to design and facilitate an information dissemination system so that information that is needed by the public can be provided in a timely manner. Our study results indicate that the government should recognize the vital role of communication and information dissemination in determining public behavior, and should pay more explicit attention to the range of factors that influence the perceptions and behaviors of the public, in order to achieve effective risk communications in emergency situations. Until the government recognizes and takes into account the complexity of factors affecting people's responses to a food crisis, government's efforts to communicate with the public and facilitate the public's access to safe and secure food supply will inevitably fall short.

Notes

1 Information available at www.pref.fukushima.lg.jp/site/portal/ps-kengai-hinansyasu.html.
2 Information available at www.maff.go.jp/j/kanbo/joho/saigai/higai_taiou/.
3 Measures: *Influential information sources* were measured by asking respondents, "With regard to information sources that influence your decisions on food safety, please indicate how much influence each information source has on your decision." A 6-point scale (0: no influence, 5: high influence) was provided for each of 19 different information sources: websites of governmental/public organizations; websites of university/research institutes; SNS/blogs; television/radio; newspapers/magazines; public relations/flyers; information provided by the local government; a newsletter published by local residents' organizations (*kairanban*); husband/partner; mother; father; parents-in-law; child(ren); siblings; relatives; neighbors; work colleagues; acquaintances; acquaintances through children; and others. *Overall connectedness to information sources* was measured by calculating individuals' mean scores of 19 information sources (M = 2.56, SD = 1.14, minimum = 0, maximum = 4.84). *Change in food-consumption behavior* was measured by asking: "Have you changed ways of purchasing or obtaining food items since the nuclear accident?" A four-point scale was provided for the respondents to choose from: "did not change at all," "did not change much," "somewhat changed," and "changed." For several analyses, the four categories were merged to two, with "did not change at all" and "did not change much" in one category, and "somewhat changed" and "changed" in the other category. *Pattern of change in food consumption* was measured by asking respondents how they changed their food-consumption

behaviors (only answered by those who indicated in response to the previous question that they either changed or somewhat changed their food-consumption behavior). Twelve statements were provided from which respondents could choose yes, no, or all that apply (e.g., "I now buy food products after checking official food screening results," "I inquire whether grocery shops or distributors conduct radiation screening on the products they carry," and "I confirm the origin of food items when shopping for groceries"). *Reasons for changing food-consumption behaviors* (answered only by those who indicated in response to the previous question that they either changed or somewhat changed their food-consumption behavior) were measured by providing nine reasons from which respondents could choose all that apply (yes/no). The categories were derived from previous survey research in Japan (Consumer Affairs Agency 2015; Food Service Industry Research Institute 2013). They are presented in the results section. *Reasons for not changing food-consumption behaviors* (answered only by those who indicated in response to the previous question that they either did not change their food-consumption behavior much or did not change at all) were measured by presenting 13 reasons for not changing food-consumption behaviors. The categories were derived from previous research (Consumer Affairs Agency 2015; Food Service Industry Research Institute 2013). They are presented in the results section. *Income* was measured by annual income of a household. Thirteen categories (ranging from "less than a million yen" to "50 million yen or higher") were provided (Mean = 6.48, Median = 6 [5–6 million yen], SD = 2.88). Education was measured by asking respondents' final educational level. Five categories were provided: "middle school graduate" (2%), "high school graduate" (35%), "technical/occupational university graduate" (31%), "university graduate" (30%), and "graduate school graduate" (2%) (Mean = 2.95, Median = 3 [technical/occupational university graduate], SD = .901). Five age categories were provided to measure age: 20–29 (21%), 30–39 (42.5%), 40–49 (33%), 50–59 (3.5%), and 60 or older (0%) (Mean = 2.19; Median = 2 [30–39], SD = .80). The low percentage of respondents in the 50s and 60s or older age groups is likely to be due to the fact that one of the filters for the survey was "having a child (children) age 14 or younger in the household." Two open-ended questions were included in the analysis to supplement and enrich the result derived from quantitative analyses. *Perceived media content* was measured by the question, "Please write detailed content or information related to food safety you have obtained from media and interpersonal information sources." *Means to reduce anxiety about food safety* was measured by asking the question, "What are the ways in which you have taken action in your everyday life to reduce anxiety related to food safety?"

4 For instance, the article published on March 21, 2011, by Nihon Keizai Shinbun criticized the government for not using the System to disseminate information to the public.

5 Details are available at www.caa.go.jp/disaster/earthquake/understanding_food_and_radiation/

6 English language information about safety standards is available at https://japan.kantei.go.jp/content/FoodandRadiation_QA.pdf.

7 Details are available at www.maff.go.jp/j/zyukyu/anpo/shishin.html.

References

Asahi Shinbun Digital. 2011. "Shutoken shinausu, mittsunokabe nenryobusoku keikakuteiden matomegai [Shortage of supplies: fuel, electricity and food] March 19.

Baur, Patrick, Christy Getz, and Jennifer Sowerwine. 2017. "Contradictions, consequences and the human toll of food safety culture." *Agriculture and Human Values*. doi:10.1007/s10460-017-9772-1.

Beck, Ulrich. 1992. *Risk Society: Towards a New Modernity*. London: Sage.

Beghin, John C. 2006. "Evolving dairy markets in Asia: Recent findings and implications." *Food Policy* 31(3): 195–200. doi: https://doi.org/10.1016/j.foodpol.2006.03.001.

Berger, Charles R. 1987. "Communicating under uncertainty." In *Interpersonal Processes: New Directions in Communication Research*. Ed. M. Roloff and G. Miller, 39–62. London: Sage.

Chino, Chieko. 2014. "Fukushimaken ni okeru nōgyono fukko wo meguru jyōkyō to kadai. [Current Status of and Pending Issues for Revitalization of Agriculture in Fukushima Prefecture.]" *Reference* 64(12): 29–53.

Consumer Affairs Agency. 2015. *Fuhyōhigai nikansuru shōhisha chōsa* [Consumer Surveys on the Perceptions of Radiation Contamination in Food]. www.caa.go.jp/safety/pdf/150930kouhyou_1.pdf.

Cvetkovich, George, and Ragnar Löfstedt. 1999. *Social Trust and the Management of Risk, Risk, Society and Policy Series*. London: Earthscan.

Dolinski, Dariusz, Wojciech Gromski, and Ewa Zawisza. 1987. "Unrealistic pessimism." *The Journal of Social Psychology* 127(5): 511–516.

Dosman, Donna, Wiktor. L. Adamowicz, and Steve E Hrudey. 2001. "Socioeconomic determinants of health- and food safety-related risk perceptions." *Risk Analysis* 21(2): 307–318.

Food Safety Commission. 2014. *Shokuhin anzen monitā kadai hōkoku: Shokuhin anzenseini kansuru ishikitōni tsuite* [Report on Perceptions of Food Safety]. Japan Food Safety Commission. www.fsc.go.jp/monitor/2608moni-kadai-kekka-yoyaku.pdf.

Food Service Industry Research Institute. 2013. *Kokusan shokuzai rikatsuyō jyōhōtēkyō shienjigyō hōkoku* [Report on the Use of Food Produced Domestically and Policy Recommendations]. https://anan-zaidan.or.jp/column/doc/kokusan1.pdf.

Goodwin, Robin, Msahito Takahashi, Shaojing Sun, and Stanley O. Gaines. 2012. "Modelling psychological responses to the Great East Japan Earthquake and nuclear incident." *PLoS ONE* 7(5): e37690. http://doi.org/10.1371/journal.pone.0037690. Retrieved May 20, 2016.

Halpern-Felsher, Bonnie L., Susan G. Millstein, Jonathan M. Ellen, Nancy E. Adler, Jeanne M. Tschann, and Michael Biehl. 2001. "The role of behavioral experience in judging risks." *Health Psychology* 20(2):120–126. http://doi.org/10.1037/0278-6133.20.2.120. Retrieved May 20, 2016.

Hamilton, Lawrence C. 1985. "Who cares about water pollution? Opinions in a small-town crisis." *Sociological Inquiry* 55(2):170–181. http://doi.org/10.1111/j.1475-682X.1985.tb00857.x. Retrieved May 20, 2016.

Harmon-Jones, Eddie, and Judson Mills. 1999. "An introduction to cognitive dissonance theory and an overview of current perspectives on the theory." In *Cognitive Dissonance: Progress on a Pivotal Theory in Social Psychology*. Ed.

E. Harmon-Jones and J. Mills, 3–21. http://content.apa.org/books/10318-001. Retrieved May 20, 2016.

Hirschburg, Peter L., Don A. Dillman, and Sandra J. Ball-Rokeach. 1986. "Media system dependency theory: Responses to the eruption of Mt. St. Helens." In *Media, Audience and Social Structure*. Ed. S. J. Ball-Rokeach and M. G. Cantor, 117–126. London: Sage.

Ho, Ming-C., Daigee Shaw, Shuyeu Lin, and Yao-Chu Chiu. 2008. "How Do Disaster Characteristics Influence Risk Perception?" *Risk Analysis* 28(3): 635–643. http://doi.org/10.1111/j.1539-6924.2008.01040.x

Huang, Hsin, Martin von Lampe, and Frank van Tongeren. 2011. "Climate change and trade in agriculture." *Food Policy* 36: S9-S13. doi: https://doi.org/10.1016/j.foodpol.2010.10.008.

Irwin, Alan, and Brian Wynne. 1996. *Misunderstanding Science? The Public Reconstruction of Science and Technology*. Cambridge: Cambridge University Press.

Johnston, Eric. 2012. "Will 3/11 prove social media watershed?" *Japan Times*, March 8. www.japantimes.co.jp/news/2012/03/08/national/will-311-prove-social-media-watershed/#.Vdq-2c6OVWg. Retrieved July 19, 2016.

Jung, Joo-Young. 2012. "Social media use and goals after the Great East Japan Earthquake." *First Monday* 17(8). doi:10.5210/fm.v17i8.4071. Retrieved July 19, 2016.

Jussaume, Raymond A., and Dean. H. Judson. 2010. "Public perceptions about food safety in the United States and Japan." *Rural Sociology* 57(2): 235–249.

Jussaume, Raymond A., Hisano Shûji, and Taniguchi Yoshimitsu. 2017. "Food safety in modern Japan." *Japanstudien* 12(1): 211–228.

doi: 10.1080/09386491.2001.11826873.

Korthals, Michiel. 2015. "Ethics of food production and consumption." In *The Oxford Handbook of Food, Politics, and Society*. Ed. R. J. Herring, 231–252. Oxford: Oxford University Press.

Kuchinskaya, Olga. 2014. *The Politics of Invisibility: Public Knowledge about Radiation Health Effects after Chernobyl*. Cambridge, MA: MIT Press.

Lin, Chung-Tung Jordan. "Demographic and socioeconomic influences on the importance of food safety in food shopping." *Agricultural and Resources Economics Review* 24(2): 190–198.

Loges, William E. 1994. "Canaries in the coal mine: Perceptions of threat and media system dependency relations." *Communication Research* 21(1): 5–23.

Masters, William A., Agnes Andersson Djurfeldt, Cornelis De Haan, Peter Hazell, Thomas Jayne, Magnus Jirström, and Thomas Reardon. 2013. "Urbanization and farm size in Asia and Africa: Implications for food security and agricultural research." *Global Food Security* 2(3): 156–165. doi: https://doi.org/10.1016/j.gfs.2013.07.002.

McMichael, Philip. 1994. *The Global Restructuring of Agro-Food Systems, Food Systems and Agrarian Change*. Ithaca, NY: Cornell University Press.

Moldovan-Johnson, Mihaela, Lourdes Martinez, Nehama Lewis, Derek Freres, and Robert C. Hornik. 2014. "The role of patient–clinician information engagement and information seeking from nonmedical channels in fruit and vegetable intake among cancer patients." *Journal of Health Communication* 19(12): 1359–1376.

Morgan, Kevin, Terry Marsden, and Jonathan Murdoch. 2006. *Worlds of Food: Place, Power, and Provenance in the Food Chain, Oxford Geographical and Environmental Studies*. Oxford: Oxford University Press.

Mou, Yi, and Carolyn A. Lin. 2014. "Communicating food safety via the social media: The role of knowledge and emotions on risk perception and prevention." *Science Communication* 36(5): 593–616.

Nakata, Tetsuya. 2003. "Shokuryō no sojyūryō kyori to sonokankyō ni oyobosu fuka ni kansuru kousatsu [The analysis of food miles and its impact on environment] Nōrinsuisan seisaku kenkyū [*Food Policy Studies*] 5: 45–59.

Nayga, Rodolfo. M. 1996. "Sociodemographic influences on consumer concern for food safety: The case of irradiation, antibiotics, hormones, and pesticides." *Applied Economic Perspectives and Policy* 18(3): 467–475.

Overbey, Katie N., Lee-Ann Jaykus, and Benjamin J. Chapman. 2017. "A systematic review of the use of social media for food safety risk communication." *Journal of Food Protection* 80(9): 1537–1549.

Reiher, Cornelia. 2017. "Food safety and consumer trust in post-Fukushima Japan." *Japan Forum* 29(1): 53–76. doi: 10.1080/09555803.2016.1227351.

Renn, Ortwin. 1992. "Risk communication: Towards a rational discourse with the public." *Journal of Hazardous Materials* 29(3): 465–519. doi: https://doi.org/10.1016/0304-3894(92)85047-5.

Rutsaert, Pieter, Zuzanna Pieniak, Áine Regan, Áine McConnon, Margôt Kuttschreuter, Mònica Lores, Natàlia Lozano et al. 2014. "Social media as a useful tool in food risk and benefit communication? A strategic orientation approach." *Food Policy* 46: 84–93.

Sano, Mariko. 2014. "Shōhisha no tachibakarano shōhisha anzen [Safety from the perspective of consumers]." *IEICE Proceedings* (Institute of Electronics, Information and Communication Engineers) 113(302): 1–4.

Slovic, Paul, Baruch Fischhoff, and Sarah Lichtenstein. 1981. "Facts and fears: Societal perception of risk." In *Advances in Consumer Research*. Ed. Kent B. Monroe, 497–502. Association for Consumer Research.

Sports Hochi. 2011. "Tokyosui panic! Kaishime zokushutsu [Tokyo water panic! People buying up water]." March 24.

Sugawara, Shin-etsu, and Kohta. Juraku. 2018. "Post-Fukushima controversy on SPEEDI system" In *The Sociotechnical Constitution of Resilience*. Ed. S. Amir, 197–224. Singapore: Palgrave Macmillan.

Sugimoto, Takashi, Tomohiro Shinozaki, Takashi Naruse, and Yuki Miyamoto. 2014. "Who was concerned about radiation, food safety, and natural disasters after the Great East Japan Earthquake and Fukushima catastrophe? A nationwide cross-sectional survey in 2012." *PLoS ONE* 9(9): e106377.

Todt, Oliver, Emilio Munoz, Marta Gonzalez, Gloria Ponce, and B. Estevez. 2009. "Consumer attitudes and the governance of food safety." *Public Understanding of Science* 18(1): 103–114. doi: 10.1177/0963662507078019.

Weinstein, Neil. D., Judith E. Lyon, Alexander J. Rothman, and L. Cuite. 2000. "Changes in perceived vulnerability following natural disaster." *Journal of Social and Clinical Psychology* 19(3): 372–395.

Wynne, Brian. 1996. "May the sheep safely graze? A reflexive review of the expert-lay knowledge divide." In *Risks, Environment and Modernity: Towards a New* Ecology. Ed. S. Lash, B. Szerszynski, and B. Wynne, 44–83. London: Sage.

Yamaguchi, Tomiko. 2016. "Scientification and social control: Defining radiation contamination in food and farms." *Science, Technology and Society* 21(1): 5–26.

Yonekura, Ritsu 2013. "Media use and information activities of foreign residents in Japan in the event of disasters." *Media Research & Studies*. NHK Broadcasting Culture Research Institute. www.nhk.or.jp/bunken/english/reports/pdf/report_13090101.pdf. Retrieved May 27, 2016.

13 Framing food insecurity and the GMO "problem" in transatlantic trade

Patricia A. Stapleton

Since the mid-1980s, the main producers of genetically modified (GM) crops have found themselves at odds with the European Union, which has largely taken a more precautionary approach to GM crop approval and cultivation. The United States—the biggest producer of GM crops in the world—has been pushing other countries to adopt its standards regarding genetically modified organisms (GMOs). This spread of American standards worries health and environmental groups, especially in Europe, who question the safety of GM food and whether GMOs actually address sustainability, hunger, and malnutrition concerns. Although the United States has seen recent shifts in consumer interest and public policymaking (at the state level, at least), the federal government continues to promote its own regulatory standards in international agreements with the goal of policy convergence—making policies more similar across different jurisdictions and levels of government.

Despite strong interest from the United States and the European Union in standardizing food safety regulations to facilitate trade, the GMO issue remains a persistent obstacle to policy convergence. At the root of this divergence are differences in how the two sides frame GMOs. Interestingly, both sides frame their stances—and subsequent policy choices—in the context of food insecurity. The United States, as a GMO proponent, argues that GMOs provide food security because their potential for higher yields increases food availability, and touts GM crops as a response to the negative impacts of climate change, such as drought. GMO opponents in the EU, however, argue that GM crops threaten food security by contributing to environmental degradation and reducing biodiversity, which in turn negatively affect the agricultural sector. Furthermore, anti-GMO organizations contend that the growing power of GMO companies and their control over seeds threaten the livelihood of small-scale farmers. These conflicting frames are the result of different technical experiences and differences in corporate power and influence, which have been translated into different political and cultural understandings of GMOs. While the definition of food insecurity at an individual level—lack of reliable access to a sufficient supply of nutritious and safe food—is not contested, the divergence between the United States and European interpretations of the causes of food insecurity and how to address them have led to trade conflicts at the international level.

In order to understand why GMOs have created this persistent roadblock to policy convergence, we must understand the differing approaches to food insecurity on both sides of the Atlantic. To do so, in this chapter I first review the divergence in technical standards, which set the course for transatlantic political and cultural differences. A review of the US regulatory stance on GMOs in contrast to the European regulatory framework helps to reveal how negotiations about GMOs serve as a proxy for institutionalized, conflicting perceptions of food insecurity, food safety, and risk. I then briefly review the development of the different regulatory regimes and obstacles to policy convergence, before turning to an overview of the Transatlantic Trade and Investment Partnership (TTIP) as an illustrative example of these obstacles. Finally, I consider how the entrenched divergence between EU and US GMO regulation might affect global trade and perpetuate food insecurity.

Regulatory evolution in the United States and European Union

US and European regulation of GMOs have diverged at several critical junctures. These points of divergence—in laboratory practices, risk assessment, regulatory structure, characterization of GMOs, experience in food safety crises, and public opinion—act as roadblocks on the path to achieving policy convergence. An exploration of the timing and nature of these obstacles helps us to understand how each side sees GMOs playing different roles in issues of food insecurity.

The initial divergence between the two trading partners can be traced to the 1970s, before GM products even entered the market, when scientists began to develop best practices for what we have come to know as modern, genetic engineering techniques (Stapleton 2016a). Due to different experiences in responding to the related risks, each side of the Atlantic developed separate oversight standards. In the UK, government-appointed committees developed safety regulations in response to an incident with smallpox pathogens that left two people dead in 1973 (Wheale and McNally 1998). These standards provided a plan for determining the level of risk involved with biotechnology experiments and required non-technical experts to be included in an advisory body for oversight (Wheale and McNally 1998). In contrast, the guidelines from the US National Institutes of Health (NIH) left risk management in the hands of the researchers themselves, with no external supervisory group to monitor compliance. By 1976, European countries had largely adopted the British standards, while the United States followed the NIH's guidelines (Wheale and McNally 1998).

Another critical divergence is the structure of the overall regulatory systems on either side of the Atlantic. The US biotechnology regulatory system remains comparatively laissez-faire in its regulatory approach for GMOs in the food supply due to the strength of the agribusiness lobby and the US desire to export food to other countries (West 2007). For example, although the United States does closely regulate genetically engineered animal products, there are fewer US regulatory standards for GM plants for human consumption (Stapleton 2016a). At the federal level, the Coordinated Framework for Regulation of Biotechnology, set in 1986, is the foundational regulatory structure for GMOs (Acosta 2014); it established that no laws were needed to regulate biotechnology products specifically (PIFB 2001). Instead, agencies already charged with oversight of food, public health, and the environment were tapped to oversee the regulation of GMOs (Acosta 2014). Individual US states might also have GMO regulations, but they cannot preempt federal law. Consequently, states generally cannot introduce stricter regulation of GMOs, although there have been a few exceptions (Acosta 2014).

Thus, the differences between the United States and the European Union include how the regulatory frameworks categorize GMOs and the right to regulate across different levels of government. The European regulatory framework explicitly rests on the categorization of GMOs as "novel" foods, addressed directly in its Regulation on Novel Foods and Novel Food Ingredients issued in 1997. This means that GMOs require a separate set of regulations in Europe, and that they are not covered by existing regulations for food, public health, and the environment as they are in the United States. GMO regulation in Europe also has several layers of governance, but member states have more power to challenge EU decisions compared to US states challenging federal regulation. Member states are bolstered by the principle of subsidiarity embedded in the structure of the European Union, which means that the EU does not take action in an area—except if it already has established exclusive competence—unless it is more effective than action taken at the national, regional, or local levels (EUR-Lex n.d.). Since 1990, the European Union has passed five major directives and four major regulations addressing GMO cultivation, processing, environmental release, marketing, and labeling.

For nearly two decades, beginning in the late 1990s, member states have challenged the European Union's regulatory framework for GMOs in several ways, creating obstacles to implementation and enforcement. In an attempt to address non-compliance, the European Union gave member states increased autonomy in GMO regulation in Directive (EU) 2015/412, which allows them to legally restrict or prohibit GMO cultivation within their own borders (Stapleton 2016b). Unlike in the United States, where federal law supersedes state law, member states have implemented additional regulations that conflict with EU legislation, such as outright bans on GMO cultivation (DG Health and Consumers n.d.). This variation in the structure of the regulatory frameworks is thus another area of divergence when comparing the United States and European approaches to GMOs.

Aside from the diverging early responses to laboratory procedures (which had lasting effects on the focus of current regulatory frameworks) and the differences in the scope of the two regulatory systems, external influences have played a significant part in the development of Europe's regulatory system. Because of more stringent European regulations, particularly in regard to labeling requirements and a ban on the importation of GMOs from 1998 to 2004, US agricultural biotech products have been refused entry into the EU marketplace. As a consequence, US leaders filed a complaint with the World

Trade Organization (WTO) in 2003, accusing Europe of "illegally restricting imports" (West 2007). The United States was successful in its petition, and the WTO ordered the European Commission to allow GM food into the European Union. Still, most member states continue to require rigid labeling criteria, creating ongoing barriers to marketplace access (Guru and Horne 1999; NFTC 2003; ITC and EC, 2016). Moreover, the WTO decision did not lead to any efforts between the two trading partners to standardize regulations; the regulatory differences were already too deeply institutionalized.

A contributing factor to the entrenchment of these early differences was a series of food safety crises and issues across Europe, which triggered a stricter risk assessment of GMOs and regulatory response than in the United States. The "mad cow" crisis of the mid- to late-1990s stands out as a major catalyst for damaging the trust that Europeans had in their government institutions to regulate food safety effectively. An array of other food safety issues—hormone-treated beef, a listeria outbreak, hoof-and-mouth epidemics, and dioxin scares—compounded the economic and political impacts of the mad cow outbreaks (Stapleton 2016b). In response to these regulatory failures and mounting public concern, the European Union embedded the precautionary principle—which takes a preventative approach to risk-taking rather than a reactive one—into food safety regulation just as US GM soy began to arrive in Europe, leading to intense public debates over the safety of GMOs (Stapleton 2016b).

In contrast to the European experience, the United States did not suffer a major food safety crisis as it developed a regulatory response to GMOs entering the market. As a result, the US Food and Drug Administration (FDA), which oversees food safety and has authority over the regulation of GM food, continues to operate from the position that GMOs are held to the same safety standards as other food products (West 2007). This approach is known as "substantial equivalence": when a GM product is genetically similar to a current "natural" crop, then the FDA considers it a natural product that can be "sold, marketed, and consumed without fear of negative impact on health or safety" (West 2007). As a rule, the FDA has not required labeling of products alerting consumers to GM content, although since 2013 there have been significant challenges at the state level in the United States regarding labeling. Europeans, on the other hand, developed and continue to maintain the position that GMOs are inherently novel, and therefore they require institutional oversight and a set of regulations separate from conventionally produced food.

In addition to the general laissez-faire approach by the US government to agricultural biotechnology regulation, the American public initially showed very little awareness of GM food (West 2007). The Pew Initiative on Food and Biotechnology (PIFB) carried out several major national opinion polls between 2001 and 2007 to determine US consumers' views on agricultural biotechnology, revealing that most Americans reported having heard relatively little about GM foods (Fink and Rodemeyer 2007). Not only did surveys show that there was a general ignorance among Americans about GMOs in food, they also revealed that most Americans were highly unaware that they had almost definitely consumed such products (Fink and Rodemeyer 2007; West 2007). PIFB concluded in 2007 that the issue of GMOs in food was not a "top of mind" issue for most US consumers (Fink and Rodemeyer 2007).

Americans' perceptions of GMOs have evolved, however, eventually arriving at a more precautious approach to GM food than in the first decades of American GMO development and cultivation. In 2014, a Pew Research Center survey revealed that a majority of Americans believe that eating GM food is "generally unsafe" (Funk and Rainie 2015). Additionally, most Americans have become "skeptical about the scientific understanding of the effects of genetically modified organisms (GMOs) on health," and "two-thirds (67%) of adults say scientists do not clearly understand the health effects of GM crops" (Funk and Rainie 2015). This shift in opinion may, in part, be the result of the public debates on a number of state-level initiatives that have been introduced to label GMOs in food, since the FDA was not requiring labels for GM content. Despite increased public attention to GMOs, understanding of the issue has not improved. An Oklahoma State University survey in 2015 found that over 80 percent of Americans support "mandatory labels on foods containing DNA," about the same number as those who support mandatory labeling of GM foods "produced with genetic engineering" (Somin 2015).

In contrast, the European Union has been labeling GM products since 1997, noting that it is the consumers' right to information, and that labeling is a tool for making an informed choice. There are also stringent EU regulations in place that require the "traceability" of GM ingredients. The last Eurobarometer on biotechnology showed that more than half of the Europeans surveyed were aware of GMOs, and that 57 percent "are not willing to support GM food" (Gaskell, Stares, Allansdottir et al. 2010). Europeans also generally have concerns about the safety of GMOs, and they have a perception that GMOs are not natural (Gaskell et al. 2010). The shift in American public opinion regarding GMOs now more closely aligns with Europeans' perception. But, whereas American public opinion has only recently arrived at declining levels of trust and acceptance, Europeans exhibited a downward trend in public opinion of GMOs from 1996 (Gaskell et al. 2010).

Although it seems as if American public opinion of GMOs is beginning to converge with that of Europeans, the regulatory regime in the United States has already been institutionalized, making it difficult for shifts in public opinion to change existing policies. Because of the nature, scope, and timing of major food safety scandals in the European Union—and the absence of comparatively significant, recent American crises—the European approach to regulating GMOs in particular, and food products more broadly, remains more precautionary than in the United States. From the outset, in their approach to GMO regulation, Europeans were responding to failures in safety regulations (in biotechnology labs as well as in the food system) that had left them vulnerable to harm. And these failures eroded public trust in the ability of political institutions, scientific experts, and corporations to protect the food supply (Stapleton 2016b). As a result, because Europeans are more wary of GM food and do not perceive it as natural, claims that GMOs are the answer to issues like food insecurity are not compelling to the European public. European food safety regulations have been built to protect consumers and evaluate a wider range of potential risks by including political, social, and economic factors in their risk assessment of GMOs, whereas American food safety regulations remain focused on assessing GMOs as they would any other food product.

The GMO "problem" in international trade negotiations

In order to overcome these institutionalized differences, the two sides would need to deliberately negotiate an agreement to solve the GMO "problem" in transatlantic trade. But these regional differences have also become clear outside of the bilateral relationship in international forums, particularly in EU–US trade disputes as seen in the 2003 US complaint to the WTO. The two sides have favored different agreements and conventions as part of efforts to create a legal foundation for their respective preferences in the international market: these include, over the years, the WTO Agreement on the Application of Sanitary and Phytosanitary Measures, the Biosafety Protocol to the United Nations Convention on Biological Diversity, and the Codex Alimentarius. Because of their divergence in the 1990s in GM product regulation, European and US actions in international forums can be interpreted as efforts to strengthen their respective positions for future WTO disputes (Murphy and Levidow 2007). This also means that the two sides approach international debates on food issues from different perspectives, making it difficult to find common ground to solve larger issues, such as how to address growing rates of food insecurity around the world.

An examination of the failed TTIP negotiations provides an illustrative case on how the GMO issue hinders transatlantic policy convergence for food products, especially regarding how the political and cultural differences between the two sides contribute to the challenges in finding common ground. In 2013, governments across the European Union gave the European Commission a mandate to negotiate TTIP with the United States. A motivating factor of TTIP was the goal of "regulatory convergence," bringing US and European standards closer together to facilitate trade. The Commission promoted TTIP with the hopes that EU and US negotiators would be able to secure a free-trade deal that would cut tariffs and regulatory barriers to trade to make access to each other's markets easier for companies on both sides (Padmanabhan 2014). Targeted industries included pharmaceuticals, cars, energy, finance, chemicals, clothing, and food and drink. An EU Commission study published in 2013 estimated that TTIP could boost the size of the EU economy by 120 billion euros (equal to 0.5% of GDP) and the US economy by 95 billion euros (or 0.4% of GDP). The Commission also promoted

TTIP as creating "the largest free trade area in the world," as well as new jobs and business opportunities (European Union n.d.). US promotion of the potential of the agreement was similar. The Office of the United States Trade Representative stated TTIP was intended to be "an ambitious and comprehensive trade agreement that significantly expands trade and investment between the United States and the European Union, increases economic growth, jobs, and international competitiveness, and addresses global issues of common concern" (USTR n.d.).

To these ends, the two sides held 15 rounds of negotiations between July 2013 and October 2016. Although the round discussions were largely kept secret, both EU and US negotiators indicated before the 11th round, held in October 2015, that they were finally going to begin to deal with agricultural concerns more directly. At the end of that round, the United States and European Union presented "a second round of goods market access offers" and were "more or less on par in terms of eventual duty reduction, with 97 percent of tariff lines expected to become duty-free at some point" (McKeon 2015). Negotiators declined to identify which issues were included in the remaining 3 percent, but due to the complexities and political challenges of addressing trade in the agricultural, textile, and automobile sectors, observers believed these areas were included (McKeon 2015).

At the end of the 13th round in March 2016, the EU Commission announced it would like to sign an agreement by spring 2017, implicitly acknowledging that an agreement would not be reached before the end of President Obama's last term in office. It became even clearer by the end of the 14th round held in July 2016 that negotiators would not be able to complete an agreement by the end of that year. Reports indicated that GMOs remained a sticking point, "both because of the complex, technical nature of approval processes and because of political sensitivities in the United States and EU" (McKeon 2016a). Even still, US Trade Representative Michael Froman and EU Trade Commissioner Cecilia Malmström publicly maintained that the two sides could achieve their "goal of reaching at least a political agreement—a nearly complete, agreed-upon framework missing technical details—by the end of 2016" (McKeon 2016a). But following Great Britain's vote to leave the European Union in June 2016, as well as the divisive presidential campaign unfolding in the United States, the prospect seemed unlikely, especially as "anti-trade sentiment continue[d] to rise on both sides of the Atlantic" (McKeon 2016a).

This was confirmed after the last round, held in October 2016, where negotiators failed to make any major breakthroughs toward an agreement. The official report indicated that talks were held on a wide range of topics for the Sanitary and Phytosanitary (SPS) Issues chapter, but that they only would "lay the foundation for continued intensive engagement" (European Commission 2016b), and not that the two sides had even approached agreement on any SPS topic. Despite the lack of a political agreement as hoped for by Froman and Malmström, the 15th round of TTIP negotiations "represented an important affirmation of the parties' will to continue negotiating in the face of an uncertain future" (McKeon 2016b).

This commitment would turn out to be brief. On November 8, 2016, Donald Trump—who argued during his campaign that international trade deals hurt US economic interests—won the presidential election (Blenkinsop 2016). On January 17, 2017, three days before Trump's inauguration, Froman and Malmström issued a joint statement asserting that "the economic and strategic rationale for an agreement between the world's two largest advanced industrial economies remains strong" (European Commission 2017). But on January 23, 2017, Trump signed an executive order pulling the United States out of the Trans-Pacific Partnership (Baker 2017). He also reiterated his promise to renegotiate the terms of the North American Free Trade Agreement (NAFTA) (Walker 2017). These anti-international trade actions were foreboding for the future of TTIP. Days later, Peter Navarro, the head of Trump's National Trade Council, declared TTIP to be "dead" (Swanson 2017).

Non-technical obstacles to policy convergence

In addition to the institutionalized technical obstacles that prevented policy convergence during the TTIP negotiations between the European Union and United States, other factors were damaging to TTIP's prospects. Although Trump's election to the US presidency and his stance on turning toward more protectionist trade policies for the United States provided a (seemingly final) death knell to TTIP, governments on either side of the Atlantic were struggling with declining levels of acceptance for TTIP in public opinion. A Pew Research Center survey from March 2014 (before the fifth round of negotiations) showed that 53 percent of Americans and 55 percent of Germans thought "TTIP will be a good thing for [the] country." But, although there was broad support for increased trade, the survey also revealed opposition on both sides to specific goals of TTIP. There was a lack of support from both the majority of Germans and Americans for "the removal of all tariffs on the transatlantic shipment of goods" and "the elimination of restrictions on transatlantic foreign investment" (Pew Research Center 2014). But more interestingly, the two sides

> sharply disagree[d] over the most ambitious TTIP objective: making transatlantic regulatory standards as similar as possible. While Americans support[ed] such an effort by a margin of 4 to 1 (76% to 18%), just 45% of Germans agree[d].
>
> (Pew Research Center 2014)

By the 13th round, public support of a deeper US–EU trade deal had plunged. A survey showed that in just two years, support for TTIP had declined from 53 percent to 18 percent among Americans and 55 percent to 17 percent among Germans (Barkin 2016).

Though German and American survey participants are not necessarily indicative of all EU and US citizens, these opinion polls did capture the declining support for a completed TTIP agreement. More specifically, the arguments raised against TTIP highlight the entrenched differences on both sides in their approach to food safety and security. Individual EU member states were becoming more vocal in their opposition to finalizing TTIP, exhibiting serious concerns that the European approach, which prioritizes consumer and environmental protections, would be diluted. An ex-Commissioner for Justice in Luxembourg, Viviane Reding, called for an end to the negotiations, stating,

> After three years of talks, we must accept that it is just not going to work. [...] Despite intense efforts on the part of the European negotiators, the United States remains firmly opposed to [...] our proposals on food security and environmental protection and even to the recognition of our protected geographical indications.
>
> (Nguyen 2016)

Other challenges to the negotiations included significant opposition in the European Union to the "investor-state dispute settlement" (ISDS) mechanism, which would have allowed companies to sue foreign governments over claims of unfair treatment and to be entitled to compensation (Ghailani and del Castillo 2015). Critics of TTIP believed that such a mechanism would undermine the power of national governments and their ability to act in the interest of their citizens (Padmanabhan 2014), particularly that such a deal could weaken EU standards in public health and the environment.

TTIP opponents also raised the issue of potential impacts of the agreement on food production and safety, "arguing that the EU has much stricter regulations on GM crops, pesticide use, and food additives than the US," and that a TTIP deal would "open the EU market to cheaper products with poorer standards" (Padmanabhan 2014). The fear was that food conglomerates would then use the ISDS mechanism to strong-arm European governments into removing food safety regulations, eroding food and environmental protections, and destabilizing food security in a race to meeting the United States at the bottom (Padmanabhan 2014). The European Commission responded to the public's concerns regarding GMO regulations, noting on its website that existing regulations of GMOs would remain untouched, and unequivocally stating, "The EU basic law on GMOs—including the European Food Safety Authority's (EFSA) safety assessment and the risk management procedure—is not up for negotiation. It will not change as a result of TTIP" (European Commission 2016a). Nevertheless, because of already uneven compliance with existing EU law, member states did not view favorably any promises that the system would remain the same.

At the same time that the reactions from member state leaders revealed the political differences between the two sides, the anti-TTIP public outcry underscored the different cultural understandings of food safety and security.

Despite the Commission's assurances, protests sprang up across Europe, often centered on the argument that TTIP would give too much power to multinational companies at the expense of consumers and workers (Copley and Reuters TV 2015). A strong undercurrent associated with these protests was the fear that the United States' cultural privileging of business over citizens would be pushed onto Europeans through globalization processes, as codified in TTIP by both the ISDS mechanism and regulatory convergence in food, public health, and the environment. TTIP opponents tapped into fears that the globalizing economy and growing power of multinational corporations would weaken consumer rights and protections and, subsequently, increase levels of food insecurity in Europe and make consumers more vulnerable to food contamination and food-borne illness outbreaks.

Politicians, journalists, and experts have identified these fears in both the UK's vote to leave the European Union in June 2016 and the election of Trump to the White House in November 2016 (Elliot 2016; Sharma 2016; Sink, Chrepa, and Talev 2016). Both events effectively halted TTIP negotiations. Barack Obama noted that "globalization combined with technology, combined with social media and constant information has disrupted people's lives," and the fear of it, coupled with suspicion of government institutions and elites, informed both Brexit and Trump's win (Sink et al. 2016). Whatever the underlying motivations for the growing populist movements in the United States and among European member states, TTIP negotiations have ended.

Free trade and food security

Trade tensions over food are not new to the international market. Prior to the final rounds of the General Agreement on Tariffs and Trade (GATT) and the creation of the WTO, agriculture was largely seen as a special case in international relationships—a commodity to be left out of free-trade agreements due to food being an integral part of national security (Clapp 2015). But there was a significant shift in that stance with the creation of the WTO in 1995, when top producers began to pressure other countries "to allow a free market in agricultural products, abandon state support for agricultural producers, and abolish measures that restrict international movement in foodstuffs" (Orford 2015). Major food exporters, like the United States, saw a benefit to having access to foreign markets for food exports, and they were also able to subsidize and protect their domestic agriculture when their markets opened to imports. In this way, "industrialized countries were able to maintain policies that supported the idea that food is different at home, while painting them as compatible with liberalization in the WTO context" (Clapp 2015). Developing countries, in contrast, were not able to maintain the same balance, as they did not have the resources available to protect domestic markets from the impacts of imports, which has led to "negative implications for food security" (Clapp 2015).

But how exactly have efforts to promote free trade affected food security? Although a number of studies engage this question, the literature does not definitively answer whether free trade helps food security or undermines it. For example, Jan Dithmer and Awudu Abdulai (2017) find that free trade impacts food security in terms of "prices, the availability of production factors, economic growth, household incomes and employment," with an overall net positive effect. Sébastien Mary (2019) counters that claim, arguing that studies showing how free trade positively affects food security do not account for the wide variation of experiences within a country. Instead, even though greater openness in food trade will increase the calories and nutrients available, "those who need it the most (living in rural areas for the most part) may not benefit from increased food trade" as they are more vulnerable to market fluctuations and less protected by government programs (Mary 2019). As a result, free-trade agreements can be a cause of food insecurity, specifically in developing countries where "a 10 percent increase in food trade openness would increase undernourishment prevalence by about 6 percent" (Mary 2019).

If the United States and the European Union had been able to come to an agreement on TTIP, there might have been repercussions for developing countries that are not well positioned to compete within a global agricultural market run by free trade principles, as well as for more economically vulnerable communities within the United States and the European Union. However, we cannot investigate a counterfactual. Instead, we must engage with how the failure of TTIP left unresolved differences in safety standards for GM food and how those differences affect food security globally.

Like free trade, the benefits from cultivating GM food are unevenly distributed, with developed countries seeing more benefits and developing countries experiencing disadvantages. Globalization, coupled with the Green Revolution (i.e., the dramatic changes in agricultural practices resulting from new advancements in pesticides, fertilizers, and crop development), has contributed to rising rates of food insecurity (McMichael 2001). As multinational corporations have been able to increase their control over the structure of the food supply through the increased use of genetically modified crops, they have undermined the ability of small farmers in both the developed and developing world to maintain their operations, making those farmers sell their goods for export rather than use them as subsistence (McMichael 2001). Moreover, as more states move to export their agricultural products, differences in food safety standards between the major powers place less-powerful states in a quandary. When EU member states set sanitary and phytosanitary standards at levels more stringent than those suggested in the Codex Alimentarius, they severely limit access to their markets (Wilson and Otsuki 2003). But fewer restrictions do not necessarily work in developing countries' favor either (Wilson and Otsuki 2003). In the case of the southern African famine in 2002, for example, the EU's moratorium on GM crops damaged the ability of other states to respond effectively to severe food insecurity (Herrick 2008). Countries like Zambia had to reject food aid from the United States because of fears that GM content would contaminate Zambian crops, which would then be denied entry to the European market (Herrick 2008). In this context, we can see how the failure of the European Union and the United States to facilitate policy convergence on safety standards for GM food has become more than just a bilateral issue (Herrick 2008).

Conclusion

So where do the United States and the European Union go from here? Even if TTIP negotiations were to reopen—which is highly unlikely—negotiators would still need to address the obstacles created by divergent GMO regulatory frameworks and understandings of the causes and solutions to food insecurity. Furthermore, actions taken during Trump's presidency (pulling out of the tentative Trans-Pacific Partnership, changes to NAFTA) will alter the landscape of international trade. Additionally, it would be highly improbable that a bilateral agreement that excludes food standards and safety can be achieved. In this way, the failed TTIP negotiations—and the role that GMOs played in preventing agreement—provide an illustrative example of how the differences in political and cultural understanding of food safety and security would require any transatlantic agreement to address the differences in risk assessment, approval, and labeling. The conflicting frames on how to address food insecurity may be the result of diverse experiences, but they now exist as their own obstacle. The United States and European Union's efforts toward policy convergence on SPS matters have ended, and GMOs will continue to remain a point of contention in international trade forums—both of which issues can distract from opportunities for the international community to respond to the continued rise of global food insecurity.

References

Acosta, Luis. 2014. *Restrictions on Genetically Modified Organisms: United States*. The Law. Washington, DC: Library of Congress Legal Reports, www.loc.gov/law/help/restrictions-on-gmos/usa.php

Baker, Peter. 2017. "Trump Abandons Trans-Pacific Partnership, Obama's Signature Trade Deal." *New York Times*, January 23. www.nytimes.com/2017/01/23/us/politics/tpp-trump-trade-nafta.html.

Barkin, Noah. 2016. "Survey Shows Plunging Public Support for TTIP in U.S. and Germany." *Reuters*, April 21. www.reuters.com/article/us-europe-usa-trade-idUSKCN0XI0AT.

Blenkinsop, Philip. 2016. "Trump Victory Could Spell Defeat for EU-U.S. Trade Deal." *Reuters*, November 9. www.reuters.com/article/us-usa-election-trade-eu-idUSKBN1342TF.

Clapp, Jennifer. 2015. "Food Security and Contested Agricultural Trade Norms." *Journal of International Law and International Relations* 11(2): 104–115.

Copley, Caroline, and Reuters TV. 2015. "Hundreds of Thousands Protest in Berlin against EU-U.S. Trade Deal." *Reuters*, October 10. www.reuters.com/article/us-trade-germany-ttip-protests-idUSKCN0S40L720151010.

DG Health and Consumers. n.d. "Rules on GMOs in the EU Ban on GMOs Cultivation." *Food and Feed Safety*. http://ec.europa.eu/food/food/biotechnology/gmo_ban_cultivation_en.htm.

Dithmer, Jan, and Awudu Abdulai. 2017. "Does Trade Openness Contribute to Food Security? A Dynamic Panel Analysis." *Food Policy* 69: 218–230. http://dx.doi.org/10.1016/j.foodpol.2017.04.008.

Elliot, Larry. 2016. "Brexit is a Rejection of Globalization." *The Guardian*, June 26. www.theguardian.com/business/2016/jun/26/brexit-is-the-rejection-of-globalisation.

EUR-Lex. n.d. "Subsidiarity." *Glossary of Summaries*. https://eur-lex.europa.eu/summary/glossary/subsidiarity.html.

European Commission. 2013. "Transatlantic Trade and Investment Partnership. The Economic Analysis Explained." http://trade.ec.europa.eu/doclib/docs/2013/september/tradoc_151787.pdf.

European Commission. 2016a. "About TTIP—Basics, Benefits, Concerns." *DG Trade*, September 6. http://ec.europa.eu/trade/policy/in-focus/ttip/about-ttip/questions-and-answers/.

European Commission. 2016b. "Report of the 15th Round of Negotiations for the Transatlantic Trade and Investment Partnership." *DG Trade*, October. http://trade.ec.europa.eu/doclib/docs/2016/october/tradoc_155027.pdf.

European Commission. 2017. "EU and US Publish TTIP State of Play Assessment." *DG Trade*, January 17. http://trade.ec.europa.eu/doclib/press/index.cfm?id=1613.

European Union. n.d. "Humanitarian Aid and Economic Prosperity." *European Union Global Strategy*. https://europa.eu/globalstrategy/en/humanitarian-aid-economic-prosperity.

Fink, W., and M. Rodemeyer. 2007. "Genetically Modified Foods: US Public Opinion Research Polls." In Dominique Brossard, James Shanahan, and T. Clint Nesbitt (eds). *The Public, the Media, and Agricultural Biotechnology*. Wallingford: CAB International.

Funk, Cary, and Lee Rainie. 2015. "Chapter 6: Public Opinion about Food." In *Americans, Politics, and Science Issues*, Pew Research Center. www.pewinternet.org/2015/07/01/chapter-6-public-opinion-about-food/#genetically-modified-foods.

Gaskell, George, Sally Stares, Agnes Allansdottir, Nick Allum, Paula Castro, Yilmaz Esmer, et al. 2010. *Europeans and Biotechnology in 2010. Winds of change? A report to the European Commission's Directorate-General for Research.* Brussels: European Commission.

Ghailani, Dalila, and Aída Ponce del Castillo. 2015. "The Transatlantic Trade and Investment Partnership (TTIP): Converging Interests and Diverging Opinions." In D. Natali and B. Vanhercke (eds). *Social Policy in the European Union: State of Play 2015*. Brussels: ETUI.

Guru, Manjula V., and James E. Horne. 1999. "Labeling of GMOs: Impact on Consumer Demand and Global Food Trade." *Nature Biotechnology* 17: 43.

Herrick, Clare. 2008. "The Southern African Famine and Genetically Modified Food Aid: The Ramifications for the United States and European Union's Trade

War." *Review of Radical Political Economics* 40: 50–66. https://doi.org/10.1177/0486613407311081.

International Trade Centre (ITC) and European Commission (EC). 2016. *Navigating Non-Tariff find Measures: Insights from a Business Survey in the European Union*. Geneva: ITC /EC. http://trade.ec.europa.eu/doclib/docs/2016/december/tradoc_155191.pdf.

Mary, Sébastien. 2019. "Hungry for Free Trade? Food Trade and Extreme Hunger in Developing Countries." *Food Security* 11: 461–477. https://doi.org/10.1007/s12571-019-00908-z.

McKeon, Michael. 2015. "TTIP Negotiations: A Summary of Round 11." *B/Brief*, November 12. www.bfna.org/publication/bbrief-ttip-negotiations-a-summary-of-round-11.

McKeon, Michael. 2016a. "Negotiators Advance TTIP Agenda but Several Key Issues Remain." *B/Brief*, July 28. www.bfna.org/publication/bbrief-negotiators-advance-ttip-agenda-but-several-key-issues-remain.

McKeon, Michael. 2016b. "TTIP Negotiators Make Small Gains, No Major Breakthroughs in Final Round before US Elections." *B/Brief*, October 21. www.bfna.org/publication/bbrief-ttip-negotiators-make-small-gains-no-major-breakthroughs-in-final-round-before-us.

McMichael, Philip. 2001. "The Impact of Globalisation, Free Trade and Technology on Food and Nutrition in the New Millennium." *Proceedings of the Nutrition Society* 60(2): 215–220. https://doi.org/10.1079/PNS200088.

Murphy, Joseph, and Les Levidow. 2007. *Governing the Transatlantic Conflict over Agricultural Biotechnology: Contending Coalitions, Trade Liberalization and Standard Setting*. London: Routledge.

National Foreign Trade Council. NFTC. 2003. *Looking Behind the Curtain: The Growth of Trade Barriers that Ignore Sound Science*. Washington, DC: NTFC.

Nguyen, Céline. 2016. "Francois Hollande Buries Hope of Concluding TTIP This Year." EurActiv.com, August 31. www.euractiv.com/section/all/news/hollande-burries-any-hope-of-ttip-conclusion-this-year/.

Office of the US Trade Representative (USTR). n.d. "T-TIP Issue-by-Issue Information Center." *Transatlantic Trade and Investment Partnership*. https://ustr.gov/trade-agreements/free-trade-agreements/transatlantic-trade-and-investment-partnership-t-tip/t-tip.

Orford, Anne. 2015. "Food Security, Free Trade, and the Battle for the State." *Journal of International Law and International Relations* 11(2): 1–67.

Padmanabhan, Leala. 2014. "TTIP: The EU-US Trade Deal Explained." *BBC News*, December 18. www.bbc.com/news/uk-politics-30493297.

Pew Initiative on Food and Biotechnology (PIFB). 2001. "Guide to U.S. Regulation of Genetically Modified Food and Agricultural Biotechnology Products." Pew Charitable Trusts. www.pewtrusts.org/~/media/legacy/uploadedfiles/www pewtrustsorg/reports/food_and_biotechnology/hhsbiotech0901pdf.pdf.

Pew Research Center. 2014. "Support in Principle for U.S.-EU Trade Pact." *Global Attitudes and Trends*, April 9. www.pewglobal.org/2014/04/09/support-in-principle-for-u-s-eu-trade-pact/.

Sharma, Ruchir. 2016. "Globalisation as We Know It Is Over—and Brexit Is the Biggest Sign Yet." *The Guardian*, July 18. www.theguardian.com/commentisfree/2016/jul/28/era-globalisation-brexit-eu-britain-economic-frustration.

Sink, Justin, Eleni Chrepa, and Margaret Talev. 2016. "Obama Says Brexit and Trump Powered by Globalization Fears." *Bloomberg*, November 15. www.bloomberg.com/politics/articles/2016-11-15/obama-says-brexit-and-trump-both-powered-by-globalization-fears, as examples.

Somin, Ilya. 2015. "Over 80 Percent of Americans Support 'Mandatory Labels on Foods Containing DNA.'" *Washington Post*, January 17. www.washingtonpost.com/news/volokh-conspiracy/wp/2015/01/17/over-80-percent-of-americans-support-mandatory-labels-on-foods-containing-dna/.

Stapleton, Patricia. 2016a. "Biotechnology Governance and Risk Regulation in the European Union." In H. F. Carey (ed.), *The Challenges of European Governance in the Age of Economic Stagnation, Immigration, and Refugees*. Lanham: Lexington Books.

Stapleton, Patricia. 2016b. "From Mad Cows to GMOs: The Side Effects of Modernization." *European Journal of Risk Regulation* 7(3): 517–531.

Swanson, Ana. 2017. "Trump's Administration Has a New Target on Trade—and It's Not China or Mexico." *Washington Post*, January 31. www.washingtonpost.com/news/wonk/wp/2017/01/31/trumps-administration-has-a-new-target-on-trade-and-its-not-china-or-mexico/.

Walker, Kristen. 2017. "Trump to Sign Executive Order on Plan to Renegotiate NAFTA with Mexico, Canada." *CNBC*, January 23. www.cnbc.com/2017/01/23/trump-to-sign-executive-order-to-renegotiate-nafta-and-intent-to-leave-tpp.html.

West, Darrell. 2007. *Biotechnology Policy across National Boundaries: The Science-Industrial Complex*. Basingstoke: Palgrave Macmillan.

Wheale, Peter, and Ruth McNally. 1998. "The Social Management of Genetic Engineering: An Introduction." In P. Wheale, R. von Schomberg, and P. Glasner (eds). *The Social Management of Genetic Engineering*. Farnham: Ashgate.

Wilson, John and Tsunehiro Otsuki. 2003. "Balancing Risk Reduction and Benefits from Trade in Setting Standards." *2020 vision briefs* 10(6), International Food Policy Research Institute (IFPRI).

Index